Praise for Dan Miller

"Dan Miller is the real deal - his experience in agile project management shines through in this comprehensive guide."

Professor Lisa Harvey-Smith
Astronomer, T.V. Presenter, MC, Speaker, Writer
CSIRO Astronomy & Space Science | CSIRO Australia

"Don't Spook the Herd! is a must read and a perfect practitioner guide, which makes sense of how Agile Project Management frameworks work effectively."

Associate Professor Ofer Zwikael
Research School of Management
ANU College of Business & Economics | ANU Australia

"Dan provides an interesting and often amusing perspective on agile, focusing on its essence: people. As he says, people are the key to successful projects."

Kathy Schwalbe, Ph.D., PMP
Author, Publisher, Consultant
Schwalbe Publishing | Minneapolis USA

"This book is written by someone who is obviously passionate about his subject area. The list of deliverables from business projects frequently includes some form of new / re-engineered business process. Dan offers some sound guidance on how agile business projects should be handled, so that the likelihood and impact of serious issues can be reduced."

John Smyrk
Owner, Sigma Management Science
Visiting Fellow, UNSW Business School | UNSW Australia

"Relationships between people are an important element in project management success. Dan is a genuine people person and as such he does a great job at highlighting the significance of socio-cultural factors to managing agile projects."

Professor Walter D Fernández, PhD
School of Information Systems and Technology Management
UNSW Business School | UNSW Australia

"Dan has written a book which is a great first step towards agility for many traditional organisations. He explores what it means to adopt an agile culture and provides a roadmap for organisations who are migrating from sequential, predictive project approaches to the learning environment needed for the 21st century."

Shane Hastie
Director of Agile Learning Programs
International Consortium for Agile

Don't Spook the Herd!

How to Get Your Agile Projects Running Smoothly

Dan Miller

Miller Productions

From the studios of Miller Productions.

GPO Box 344
Canberra ACT 2601
Australia

https://miller.productions/

First edition

Cataloguing-in-Publication entry is available from the National Library of Australia:
http://catalogue.nla.gov.au/

Title: Don't Spook the Herd! - How to Get Your Agile Projects Running Smoothly

Author: Miller, Dan (1977-)

ISBNs: 978-0-6481154-1-0 (Paperback)
978-0-6481154-2-7 (eBook - EPUB)
978-0-6481154-3-4 (eBook - Kindle)

Dedicated to Kristin, Amelia, Zoe & Eric.

Thanks for being agile and supporting me as I switched priorities to write this book. I love you always and forever!

Contents

Foreword

Projects are all around us and in an increasing number and level of strategic importance. They help business grow and improve. However, projects have also become more complex over time and the expectations from project managers grow. If in the past a project was considered successful when their outputs satisfied the iron triangle (time, cost, scope), now the projects also need to deliver benefits.

No surprise then that new project management methodologies have appeared to provide an effective response to these challenges. Above all, Agile Project Management appears to be the most successful one, judging by its acceptance in practice all around the world.

Transforming the software development industry, this methodology provides flexibility and speed that allow projects delivering better results. However, as this book correctly states, Agile Project Management is a generic approach that can fit many other types of projects and industries.

Don't Spook the Herd! is a practical guide to Agile Project Management. The clear summary of various Agile approaches and simple examples make the book an excellent introduction to the world of Agile Project Management.

Don't Spook the Herd! provides an excellent emphasis on crucial project management areas. Whereas project management methodologies tend to focus on the "hard" processes and tools, this book discusses the "herd", in particular the important factors of people, culture, stakeholders and leadership, to name a few.

Further, *Don't Spook the Herd!* provides a unique link between Agile Project Management and the implementation of organizational strategy through and realisation of project benefits.

Don't Spook the Herd! is a must read and a perfect practitioner guide, which makes sense of how Agile Project Management frameworks work effectively.

Associate Professor Ofer Zwikael
Research School of Management
ANU College of Business & Economics | ANU Australia

Everything I Thought I Knew About Project Management...

I used to think that the only *real* way to manage projects was through predictive planning. I believed in rote application of various textbook approaches and surgical-like command and control.

But that was during the early days in my career. As is the case with most people starting out, this line of thought was due to my limited experience. It also didn't help that I followed a broad and scattered approach to the sources of information used for guidance.

Then, sometime in the late 1990's, I came across various techniques that were the pre-cursor to Agile Project Management.

For a few years I tinkered around with the idea of agile. I got to see it applied *in-name-only*, i.e. not well at all. There were hints to me that this approach had potential, but I still wasn't convinced.

I used to think that this agile approach was just another industry buzz that had sprung up with good marketing behind it.

But then, about 10 years ago, I had the good fortune to work on a project with a group of people that really wanted to do agile well. With the guidance of a few friendly and respected people, the culture of this particular group was extremely positive. They would always support each other. They accepted their mistakes and used them as opportunities to learn and improve together. Importantly, they believed in what they were doing.

As a result, the rest of the organisation believed in this group and supported them in their quest to succeed with agile project management.

After collaborating with that group for many years I am now convinced that under the right conditions, agile project management works well.

We completed several projects together, successfully.

We were nominated for, and won, several industry awards and organisational achievement medals. Even today, people often refer to those projects in such a positive reminiscent manner.

That's enough about my experience for the moment.

How about you?

Let's begin with a question: How do you think projects should be managed?

Do you feel there is a certain way that works for you? Are you comfortable with a method based on your own experiences, intuition and feelings?

Perhaps you favour evidence-based results and prefer to align your choices with approaches that have proven to deliver success. Maybe you are somewhat sceptical about the *latest and greatest* ideas.

Alternatively, you might be the curious type and like to try out new things. Do you believe it depends on the type of project you are running and in which industry you are operating? Or do you believe it is independent, with the management style being agnostic of the project type and industry?

Next question: What stage are you in your career?

Do you have a few runs on the board, say mid-career? Maybe you're just starting out having only been in the workforce for a short time. Even earlier?

Perhaps you are progressing through your own education and training prior to really getting out there in your chosen industry.

Alternatively, you may be a seasoned professional that has completed a lot and can, sincerely, compliment much of what I am about to share with your own experience.

Last question: What do you know about agile project management?

Are you working with an organisation that runs projects using agile and you already know a thing or two? Perhaps you've only heard about it but haven't really experienced it in practice.

Have you worked somewhere that said they applied agile project management concepts but for some reason it turned out to be a disaster? Alternatively, you may have seen more positive results with agile project management.

Regardless of how you answered these questions, I truly believe there is something in this book for you.

My goal is not to convince you of anything or win any arguments. Instead, the primary objective of this book is to share with you the numerous ideas, successes, failures, strategies, techniques, tools, tips and tactics I have experienced.

I have been lucky enough to have over twenty consecutive years of working in and managing projects. For around fifteen of those, most of the projects I have been involved with used agile.

There is plenty covered in this book. The topics range from concepts, tools and processes to people, feelings and motivation.

Good Processes and Tools Make a Difference

A good set of tools in the right hands makes a big difference to productivity. A practical guide like this wouldn't be much use without coverage of the tools of the trade that will help make your agile projects run smoothly.

In this book, I provide several examples of various tools, techniques and processes you can use to help get the most out of your agile projects. I introduce some of the more common mechanisms and help you get up and running with them.

You will also find a few hidden treasures that I have created and used in real agile projects over the years.

There is an important insight to keep in mind when it comes to agile projects. While good tools do indeed help, they are not the only thing you should focus your attention on. There is something else that is way more important.

Pay Attention to the People

An underlying theme that you will see repeated throughout the book is the importance of people in agile projects.

When the people involved in an agile project are comfortable and feel confident that activities are being managed effectively, things will run smoothly.

A project is like a graceful herd of deer.

When the herd is at ease they can be the most fantastic group. They are capable of amazing feats of speed, coordination and agility.

However, when they are freaked out for some reason, chaos sets in, and they end up all over the place.

This is much like an agile project.

In other words, *Don't Spook the Herd!*

PART I

DEFINITIONS

Chapter 1
Let's Begin on The Same Page

What are we talking about here? What exactly do we mean when we say agile?

Whenever you start out on a journey together with someone, it is always good to know where you are going first. Regardless of how familiar you are with agile project management it is possible you may have your own understanding and opinion toward it.

Let's begin our journey through this book by providing a basic definition of agile. The Oxford Dictionary defines the term agile[i] as follows:

Able to move quickly and easily

Relating to or denoting a method of project management, used especially for software development, that is characterized by the division of tasks into short phases of work and frequent reassessment and adaptation of plans

Agile methods replace high-level design with frequent redesign

That's a pretty good starting point to get a handle on what agile is and what we're going to be talking about in this book. Next, let's break this down into its parts and discuss each one.

A Method of Project Management

Agile is a way of running a project. It covers various things including:

- roles you have for people on the project;
- ways you go about gathering requirements and defining the project's outputs;
- how you schedule the work;
- what approaches you use to monitor and control the work;
- ceremonies you follow for various interactions between people;
- steps you take to ensure the priorities of the project match what key people require;
- methods for demonstrating progress and accepting delivered output.

Used Especially for Software Development

This is true. Professionals within the software development industry have embraced agile as a way of working.

There are many reasons for this which we won't cover here other than to say that agile is a close match that suits the characteristics of software development projects.

There is a famous public declaration that was put together by a group of prominent and leading figures in the software industry back in 2001.

The declaration was named the Manifesto for Agile Software Development[ii] and states the following:

We are uncovering better ways of developing software by doing it and helping others do it. Through this work we have come to value:

Individuals and interactions over processes and tools

Working software over comprehensive documentation

Customer collaboration over contract negotiation

Responding to change over following a plan

That is, while there is value in the items on the right,
we value the items on the left more.

This manifesto became the clarion call and guide for thousands, if not millions, of software industry professionals from then onwards.

Around this time, a significant body of knowledge grew around the application of applying agile management to software development.

The main reason that people associate agile with software, in my opinion, is that the software industry was where agile really became prolific and pervasively applied. In other words, the folks in the software industry got in first and made it their own. However, it is very important at this stage of the book to clarify something.

Agile is not just for software development projects.

This book covers agile in its generic application on any type of project in any industry wherever there is a good fit. In fact, you could easily re-apply the Agile Manifesto to be applicable to any context with a minor modification of just three words.

We are uncovering better ways of developing ~~software~~
[project outputs] by doing it and helping others do it.
Through this work we have come to value:

Individuals and interactions over processes and tools

~~Working software~~ [Completed project outputs] over comprehensive documentation

Customer collaboration over contract negotiation

Responding to change over following a plan

That is, while there is value in the items on the right, we value the items on the left more.

Many different types of projects in many different industries are good candidates for working with agile. It comes down to the characteristics of the project itself. In Chapter 3, I describe some of these characteristics and discuss why they matter in relation to agile.

Division of Tasks into Short Phases of Work

A good way to visualise this part of the agile definition is to think about an example that most people would be familiar with. We then consider that example in terms of time blocks, a.k.a. iterations.

Let's use a simple example: hosting a dinner party.

This example we are using is deliberately simple. The idea is to help get you comfortable with the basics of how agile works. As you go through it, imagine how you might apply some of these agile ideas to a more complicated real-world project.

The main output of this example is an event. It includes confirmed guests attending, a set of cooked and served dishes, a table and seats to serve a meal, followed by cleaned equipment and a tidied venue.

One way to view this project in a non-agile manner would be to break it into two phases:

- Phase 1 – Design: What's on the menu? How it will be cooked? Where it will be served? Who will be invited?
- Phase 2 – Work: confirm attendance, gather ingredients, gather cooking tools, confirm table and seating availability, confirm dishware and utensils, prepare ingredients (i.e. washing, chopping, etc.), set up the table, cook, plate, serve and lastly clean.

If you were to run this same project in an agile manner instead, you might split things into a higher number of shorter time blocks. For example, you may have something like the following, with equal blocks of time or iterations:

- Iteration 1 – decide on menu and serving location.
- Iteration 2 – send invitations, confirm attendance.
- Iteration 3 – gather ingredients and confirm quality.
- Iteration 4 – prepare table, seating, dishware and utensils.
- Iteration 5 – prepare ingredients and gather tools.
- Iteration 6 – cook.
- Iteration 7 – plate and serve.
- Iteration 8 – clear and clean.

Frequent Reassessment and Plan Adaptation

Keeping with the example of the dinner party, we can pick up a few risks.

What if you find that the ingredients you need are not available? What happens if you get half way through the project and find out someone has a unique dietary requirement? What if you find out your team just doesn't have the skills necessary to cook that outrageously complicated dish you have selected? Etc…

In the non-agile project above, by the time you got through phase 2, it would probably be too late to make any adjustments. Or at the very least, it could turn out to be a major issue with significant impact if you needed to change certain things later.

Imagine finding out after you had finished cooking that some of the guests had a deadly nut allergy and wouldn't be able to eat your satay chicken entrée or your massaman beef with peanut curry? What a disaster!

In an agile project, we provide opportunities along the way for the key stakeholders of the project to check-in and review the plans and outputs. During these assessments, the stakeholders can confirm that the outputs being produced match their requirements and constraints. They may also adjust targets and plans if necessary.

Running that dinner party example with agile, you would have several opportunities to assess and adjust along the way. At the end of iteration 1, you could make a call on whether the location and menu were suitable for the type of dinner party you wanted. At the end of iteration 2, you would have better knowledge about the number of people coming and what their dietary requirements were. Following iteration 3, you could adjust the menu if certain ingredients were not available or were of low quality. And so on.

Replace High-Level Design with Frequent Redesign

This part of the definition is the driving force behind agile.

In non-agile projects, you often see a phase toward the beginning where a large proportion of the work goes through gradually increasing levels of design. The outputs being produced by the project are designed up-front, so stakeholders can visualise how they will turn out.

In non-agile projects, the resources consumed in the up-front design process usually limit the amount of re-work and re-design that is possible. Because of this, the people on the project attempt to predict as much as they can early on.

On the other hand, agile projects put less emphasis on up-front design.

The people on the project focus more on producing outputs and regularly providing opportunities for stakeholders to review the work. The goal is to ensure the outputs are a good fit to their requirements.

In an agile project, there is less emphasis on up-front design and estimation. There is more emphasis on regular planning, production and delivery of outputs. Progress is supported with reviews by the end users of those outputs. This is known as the *Plan-Output-Review* cycle.

The agile focus on plan-output-review cycles enables progress to be gained in the real-world with tangible outputs. Compare this to the non-agile approach of spending heavy amounts of time up front in the conceptual world of thoughts, ideas, drawings and specifications.

The Essence

I used to think the lack of up-front design, documentation and predictive planning in agile was a recipe for chaos at-best and disaster most-likely. "Are you kidding me, this will never work!" I would scream internally to myself.

You too could be forgiven for thinking the agile process is suggesting we stumble ahead blindly on our projects and hope everything just works out.

When I began working on agile projects I started to realise that wasn't the case at all. I noticed that there was something going on that meant we did produce things that our customers and stakeholders wanted. These projects generated the target outcomes and benefits that the projects were initially started for.

Catching that first glimpse of a gem in the rough was enough to get me digging further into it. After decades of experience working-on projects, as well as researching and teaching project management, my current thinking about agile is as follows:

The iterative plan-output-review structure in a well-functioning agile project ensures everyone has sufficient information available at the right times.

Self-organising teams can focus on delivering outputs that match what the project stakeholders need.

This is my way of boiling a successful agile project down to its essence. Let's break it down into its parts.

These are the basic principles I see as core to agile:

- An iterative plan-output-review structure.
- Sufficient information available at the right times.
- Self-organising teams.
- A focus on delivering outputs.
- Delivering a match to what the project stakeholders need.

The Right Fit

The last concept we consider in this chapter is *fitness for purpose.* For agile to work effectively, there needs to be a good fit across four areas:

- The project needs to have certain characteristics, i.e. it needs to be a particular type of project.
- The project environment needs to facilitate and enable the agile approach.
- The organisational culture needs to be supportive of incremental progress and adjustments based on results.
- The people involved must be respected, and sometimes nurtured, in relation to the way the project is managed so they can develop their own trust in the process.

With that, our initial definitions are complete. We now have a shared understanding of the fundamental principles of agile project management.

Don't get too comfortable though, we're about to launch knee-deep into a common source of confusion in the agile world. Get ready to be bamboozled by the multiple flavours on the agile menu.

Chapter 2
Flavours of Agile

Have you ever walked up to an ice-cream shop that has several rows of different flavours and been baffled by choice?

> *Hmmm, chilli-chocolate, what a strange combination. I wonder what blue-surf is? Since when did caramel and salt become a thing? Wow, butterscotch-banana-bunch! Double-choc-mocha-almond-fudge, oh my. This is all too hard, just give me vanilla.*

You would be forgiven for doing the same thing with agile. A quick search these days for the phrase *agile methodologies* returns a big list.

The names of various agile methodologies available may as well be ice-cream flavours when you first look at them. Some of them sound like they were invented to sell to unsuspecting tourists.

What is your *vanilla* when it comes to managing projects? What is the tried and true methodology you are most comfortable with? Whatever it is, you have obviously purchased this book because you are an adventurous type and you would like to try a new flavour.

In this chapter, I tantalise your taste buds by taking you on a tour of the various flavours of agile. We'll find some are nice and simple, some have many different fillings, and one or two really are just frozen custard masquerading as ice-cream.

Grab your sample spoons and let's go!

Aficionados and Proliferating Sub-Genres

It is human nature for people to want to classify things. This is especially true when the people involved are enthusiasts.

You only need to look through the sections in a record store to see what I'm talking about. Music enthusiasts classify things down into genres such as rock, pop, classical, blues, electronic, metal, etc. Aficionados of each genre then split things down even further.

For example, within the genre of electronic music you get the sub-genres such as ambient, breakbeat, dub, trip-hop, techno, trance, industrial, gabber, hardcore, happy-hardcore, jungle, garage, house, etc. But it doesn't stop there. Within the house sub-genre, you get Chicago-House, French-house, hard-house, deep-house, acid-house, progressive-house, tech-house, electro-house. It just keeps going!

Why does this sort of classification and sub-classification happen? Basically, it is the result of people taking an existing idea and improving on it or evolving it to suit a specific purpose. It is a by-product of innovation. In the majority of cases, you would have to agree that innovation is a good thing.

The classification of a style down into sub-genres is a by-product of innovation. It's a good thing. It means ideas are progressing and people are developing specialist skills.

Innovation such as this has occurred with agile project management. The category of agile has existed as its own genre within the broader project management space for quite a while.

Over time, enthusiasts and aficionados have arisen and evolved agile into various spin-offs. Several flavours of agile have been created.

As you will see in just a moment, many of them are heavily coupled with software development since this is where agile gained early traction. However, like I mentioned previously, agile isn't just for software projects.

The Menu

At the time of writing, the following flavours of agile were known to be in existence:

- Scrum;
- Kanban;
- Scrumban;
- Crystal Methods;
- Extreme Programming (XP);
- Agile Unified Process (AUP);
- Disciplined Agile Delivery (DAD);
- Lean Software Development (LSD);
- Feature Driven Development (FDD);
- Rapid application development (RAD);
- Adaptive Software Development (ASD);
- Dynamic Systems Development Method (DSDM).

There are probably other obscure sub-genres out there, I expect. However, the point of this chapter is not to go into the minutiae but rather to give you a general overview.

For the purposes of this book, I think there is more value in people becoming aware of the range of agile methodologies and being able to distinguish between them.

The existence of this list is possibly one of the reasons you purchased this book. You may have heard about some of the following sub-genres of agile and wondered what the heck was going on.

I mean, just look at it. It's acronym city!

Some of those names sound like moves out of a beginner's guide to wrestling!

If this is your first-time hearing some of those names, I wouldn't be surprised if you were scrunching up your face right about now and thinking you had landed on some other planet.

Anyway, let's go through each of them and try to alleviate the confusion.

Today's Specials

Ladies and gentlemen, thank you for dining with us this evening. Allow me to take you through our blackboard specials. We have a lovely range of methodologies hand-crafted on-site using in-season ingredients. But first, a short story on the origin of our recipes.

A large proportion of agile methodologies have a strong connection with software projects. I used to think the reason behind it was that software professionals like building new systems. I found out, however, there is a more sombre reason: failure.

As you will soon see, the 1990s was a time of explosion in terms of new project management processes and methodologies in the software industry. A lot of this was due to the historical failures observed in the industry up to that point.

You see, the software industry was in its infancy around that time. For several decades up until about the 1970s, software projects were limited to a relatively small number of well-resourced organisations.

As the 1980s rolled around, more and more organisations began to dabble in software projects of their own.

During that time, well-meaning managers attempted to apply age-old project management techniques to these new software projects.

Unfortunately, as many organisations found out, trying to use the same approach for a software project that you would on an engineering or construction project just didn't work. As it turns out, there is a significant degree of complexity and unpredictability when it comes to building software.

The old predictive planning approaches to project management just didn't stand a chance. Unbelievably large amounts of money were lost on unsuccessful software projects in those days.

The growth in software project management methodologies, including agile methodologies, was a response to this situation.

The set of methodologies described next are the main agile sub-genres that proliferated during this period. Some have withered, while others continue to flourish to this day.

Scrum

I have deliberately started the coverage of agile sub-genres with Scrum. I have also given it the most thorough description out of all the agile sub-genres covered here. The reason will become apparent in the next section. For now, I invite you to pay close attention to the details of Scrum I am about to describe.

Scrum is a project management methodology that can be applied to many different industries. It provides a framework and structure that guides the main activities a team undertakes to reach the goal of producing the outputs of the project.

The word scrum comes from the sport of rugby. It was conceived back in 1986 by a pair of researchers, Hirotaka Takeuchi and Ikujiro Nonaka. In an article they wrote for the Harvard Business Review, Takeuchi and Nonaka introduced a new concept of product development being undertaken like a game of rugby:

> *The process is performed by one cross-functional team across multiple overlapping phases. The team tries to go the distance as a unit, passing the ball back and forth.*

The approach was popularised by Ken Schwaber and Jeff Sutherland who used it in their own companies to support projects that produced software products.

In rugby, the main feature of the sport is the game. Within each game, players focus on the ball and whoever has it. The point is for a team to combine a set number of plays and get the ball to their end of the field to score.

Within each game, there will be a set of plays the team would like to try and complete successfully. This set is sometimes known as the playbook.

Opposition players regularly tackle the person with the ball. When there is a tackle, a large group of players from both sides form a pack around that tackle. This pack that forms is called a scrum.

When a scrum forms, everyone in the pack locks heads and shoulders together and rummages around until finally the ball pops out the back of the pack.

The ball gets passed around and each player that has it sprints very fast to try and get down to their end of the field and score.

Each time a team scores, there is a break in the game where players can pause, have a chat about what just happened, and then plan out their next set of plays. After the pause, the game resets back to the middle. The game ends after a set period of time.

The analogy goes like this. A project is like a game of rugby. The project team members are like the players in the game.

The outputs the team members produce are like the plays in the game and the score at the end. These outputs are known in Scrum as *user stories.*

Just like the playbook in a game of rugby, there are a set of outputs in a project that the team would like to try and complete. This set is known as the *backlog.*

The period during which the team works to produce outputs is known as a *sprint*, i.e. when players are passing the ball between each other and running towards their end to score. The project is divided into a set number of sprints. At the start of a sprint, the team gets together to *plan* and *estimate* what they will work on.

During a sprint, the project team members regularly get together in their own group, i.e. a *scrum*, to focus on producing outputs.

In a rugby match, the team focuses on each individual play in isolation. The aim is for the whole team to work together to execute each play successfully to completion.

Sometimes, there are opportunities for the team to split into separate parts, i.e. forwards and backs, and run plays simultaneously. For example, the forwards could play a blocking role while the backs get ready to spread out and run.

Similarly, a Scrum project team focuses on small, distinct units of work. Each unit of work, i.e. output, is known as a *story*. Each story could provide value in isolation to end users either as a standalone product or as a new feature/function of an existing product. In a game of rugby, the team will try to complete a number of set plays until they score. Similarly, a Scrum team will try to complete a set number of stories each sprint.

After each sprint, the team delivers their outputs. They do this with the project stakeholders in a session known as a *demonstration*. The demonstration, or demo, is like the moment in a rugby match where the team scores. The scoring team then runs around on the field patting each other on the back while the supporters cheer.

Following delivery, the project team pauses to reflect on things that went well in the last sprint. They also consider work practices that could be improved in future sprints. This pause and reflection is known as a *retrospective*. After the retrospective, the team resets and starts on a new sprint.

There are a few set roles within a Scrum project. That is, nominated positions with associated responsibilities that are assigned to individuals.

In a game of rugby, there is a referee on the field that helps keep the game moving and the players focussed on staying within the rules. Just like the referee in a game of rugby, a Scrum project has a role known as the *Scrum Master*. The role of the scrum master is to keep the project moving and to guide the project team members to stick with the overall Scrum processes.

A rugby team usually has many stakeholders such as supporters (a.k.a. fans), administration staff, coaching staff, trainers, owners, sponsors, etc. The team on the field is playing to deliver a result for all the stakeholders, i.e. set plays that produce scores and a win overall.

They also undertake activities off the field to meet the needs of stakeholders. For example, the team trains regularly, they appear in media interviews, they undertake promotion activities and they perform community services etc.

It is the job of the club CEO to represent the views of all the stakeholders and ensure the team focuses on the activities to support those requirements.

In a Scrum project, the role that represents the views of the stakeholders is known as the *Product Owner*. The product owner pays attention to ensuring the team works on outputs that meet the needs of end users, and that the outputs can deliver the target outcomes and benefits of all the other project stakeholders.

Lastly, in rugby there is a coach that usually sits on the sidelines and observes while the team on the field does their thing. Occasionally the coach steps in to provide guidance or to resolve issues that might occur. The same is true in a Scrum project. A project manager can support the team by observing and ensuring momentum is maintained. If there is ever an issue on the project they can help resolve it. The project manager can occasionally step in to provide coaching and guidance to the team.

Kanban

Kanban is an approach to project delivery based on three principles:

1. Visualise the work. The most basic form of this is a board with three columns showing work to-do, in-progress, and done.
2. Reduce bottlenecks. Avoid overload by limiting the amount of work in progress.
3. Deliver continuously. Team members pull the next item to work on from the top of the prioritised to-do list.

Kanban is the oldest of all the methodologies and traces its heritage back to the work of Toyota in the late 1940s. It is based on the concepts of Just-In-Time delivery and Lean manufacturing.

Kanban is more suited to day-to-day operations rather than one-off projects. The idea behind it is that work continues to come down the line and is placed in a *backlog*. Team members pick up items from the backlog, work on them and then complete the workflow through to done.

New work keeps coming and everyone keeps things flowing smoothly by restricting the amount of work that is underway concurrently at any given time.

Kanban works on projects where there is a high degree of uncertainty about what work is required, and where there are frequent priority changes.

Like Scrum, Kanban can be applied to any industry. It is very lightweight and quite easy to implement.

Scrumban

Scrumban is one of those hybrids that is inevitable when two similar styles of things become popular. Just like the combination of a croissant and a donut is a cronut, Scrumban combines the best bits of each approach into one. although, if I'm being completely honest, I'm not so sure I'd be that interested in trying a cronut.

The idea is that teams use Scrum as the framework for the management of the activities and interactions for the project. They then also use Kanban as the way in which they carry out the work of producing the outputs of the project.

Some of the Scrum formalities remain active such as planning, demos, retrospectives and daily stand-ups. However, they are a little more lightweight and some are triggered differently.

Iterations, or sprints, are optional in Scrumban. If they are desired, they are kept much shorter than Scrum. Instead, team members continuously work through the backlog and limit the amount of work-in-progress. This helps the team to adapt and respond more easily to changing priorities.

Regarding prioritisation, stakeholders can continuously state their desires and change their mind at any time. They are not limited to fixed priorities per-iteration.

Rather than having set planning sessions at the beginning of each iteration, planning is triggered when the number of items remaining in the to-do or up-next state drops down below a certain threshold. The planning tops-up the number of items ready-to-start.

Scrumban lends itself to scenarios where there are smaller teams, less complex outputs, work is event-driven rather than plan-driven, and there is a need to respond to changing priorities.

Using Scrumban can also help organisations transition from the relatively less formal and unstructured Kanban to the more structured and formalised approach of Scrum.

As put by Bernie Thompson, the person credited for first documenting this hybrid approach, Scrumban gives you the flexibility of Kanban and the scaffolding of Scrum[iii].

Crystal Methods

The Crystal Methods, no it's not a band (you're in the wrong place), is a group of lightweight methodologies in the agile family. It is more commonly known as Crystal.

Crystal Methods are to agile project management as a concept album is to the music world. They are a collection of properties, strategies and techniques with a similar underlying theme in relation to running successful projects.

Crystal originated in the mid-1990s from Canadian computer scientist Alistair Cockburn. Cockburn was at the time working with IBM on research relating to developing a new methodology for technology projects[iv]. His research highlighted cases of projects that didn't follow formal methodologies but were each successful. Cockburn went looking for common core elements among these projects that may have contributed to the success.

Each of the individual methods in the Crystal family were colour coded, e.g. Clear, Yellow, Orange. The classification was designed to note the applicability to projects based on size and criticality, i.e. small & trivial through to highly complex, mission critical and potentially life-endangering.

Seven common properties were identified:

- frequent delivery;
- reflective improvement;
- close or osmotic communication;
- personal safety;
- focus;
- easy access to expert users;
- technical environment with automated tests, configuration management, and frequent integration.

The name Crystal was a play on the idea that the faces of a gemstone, i.e. a crystal, are all just windows or views into a common core. Each of those projects that Cockburn observed in his research were just a different view on a common set of underlying success factors.

Extreme Programming

Extreme Programming, or XP, started in the mid-1990s in the software industry. XP, like Scrum, has quite a structured approach to how a project is run.

XP goes a bit further and specifies details of how the project team should create their outputs. In other words, it is quite prescriptive in how the team is expected to undertake their work.

The essence within XP is a combination of values, practices and rules.

Agile Unified Process

Remember how I mentioned that the 1990s was a time that saw large growth in various agile project management methodologies. It appears coming up with new agile practices was the in thing.

Given this, what do you think normally happens when something becomes popular? That's right, commercial organisations see an opportunity and they throw their hat in the ring.

During the mid-1990s, the Rational Software Corporation obtained a software process asset during a strategic acquisition. It is assumed that one of the goals of acquiring this process was to provide additional value to customers of its associated products and consulting services. It named this process the Rational Unified Process, or RUP. The commercial nature of RUP meant it is integrated closely to Rational's suite of products and its consulting practices.

In a nutshell, RUP was a tailorable process that guided software development. It is based fundamentally on another underlying approach known as the Unified Process. It was also described by some people as a bit complicated.

As is the case when something useful is a bit complicated, people aim to simplify and customise it.

Agile Unified Process, or AUP, is exactly that. Arising in 2002, AUP is a simplified version of RUP, customised for use with agile project management. In the words of AUP's creator[v]:

It describes a simple, easy to understand approach to developing business application software using agile techniques and concepts yet still remaining true to RUP

AUP is an iterative and incremental process built around a set of phases, disciplines and philosophies.

Disciplined Agile Delivery

Disciplined Agile Delivery, or DAD, is relatively new on the scene. Amazingly, it wasn't developed in the 1990s! In the words of its creators[vi]:

Disciplined Agile Delivery (DAD) is a people-first, learning-oriented hybrid agile approach to IT solution delivery. It has a risk-value delivery lifecycle, is goal-driven, is enterprise aware, and is scalable.

DAD is aimed at IT projects and combines some of the strategies of various other project management methodologies. DAD accepts the fact that many organisations like to implement their own custom versions of work methods. DAD takes a non-prescriptive, goals-driven approach.

The way DAD is defined lets users tailor the methodology to match their project needs. Another focus of DAD is attempting to address the issue of not working so well on larger complex projects, something other methodologies tend to suffer from.

Lean Software Development

Lean Software Development, or LSD (I'm not so sure they thought that acronym through) is, in the words of its creators[vii]:

A toolkit for translating widely accepted lean principles into effective, agile practices that fit your unique environment.

In the context of manufacturing products, the term Lean refers to an approach that originated from the Toyota Production System in the 1990s. Lean Software Development is an application of the principles of Lean to the software industry.

Lean Software Development is what you could call a cousin of agile. That is, rather than being a specific project management method, Lean Software Development encompasses a set of principles that support agile.

Of course, as the name suggests, Lean Software Development is focussed on building software projects. However, if you take a step back you could see how the principles and tools could be applied generically to any type of project that produces a series of outputs.

Feature Driven Development

Feature Driven Development, or FDD, is an agile method of developing products where the features of each product are the primary focus.

A feature in this context is simply some function or capability of the product.

Features are based on what is useful for the client of the project. That is, a feature is something that provides value to the end users of the product.

Features are expressed using a specific terminology that has three parts. First there is the action that the feature will perform, then there is the result that is to be produced, and last there is the object which the action and result are applicable to.

For example, imagine you were producing a solution that helped you manage student enrolments for a school or university. One of the features could be to list the students enrolled in a class. The action there is to list, the result is the set of students enrolled and the applicable object is the class.

FDD is guided by models that help the project team understand the domain and context of the system or product(s) being produced in the project.

Rapid Application Development

Rapid Application Development, or RAD, arose in the software industry during the 1980s and 1990s. RAD was a response to the slow and artefact-heavy methods such as Waterfall that seemed to be leading to regular project failures.

The key ideas behind RAD are:

- minimal planning;
- up-front requirements gathering;
- regular interaction with users of the project's outputs;
- solution designs based on models;
- parallel development of components;
- verification using working prototypes;
- rapid and iterative build, refine cycle;
- less formality in stakeholder reviews and communications;
- less pre-defined processes on running the project;
- re-use and integration of components and prototypes;
- cutover from development to production system.

Adaptive Software Development

Adaptive Software Development, or ASD, is another method that began in the software industry. ASD grew out of the Rapid Application Development world.

ASD arose from the observation of limited success with project management approaches that relied on predictive planning.

The premise of ASD is that projects have a high degree of unpredictability and rather than this being a constraint, it is embraced as an enabler of progress.

The basic characteristics[viii] of ASD are that it is:

- mission focused;
- feature based;
- iterative;
- time-boxed;
- risk driven;
- change tolerant.

Continuous change in the project and in the outputs of the project are normal.

The old idea of a static plan-design-build lifecycle for a project and its products is replaced by a speculate-collaborate-learn lifecycle.

There is flexibility as to when any of the phases take place. While a cycle is recommended, the phases don't always need to happen strictly in a certain order. If kicking off one phase early will help the other phases, then the project should do so.

The idea is that the project framework, the team and the stakeholders can adapt and evolve as they make progress toward the overall goals.

Dynamic Systems Development Method

Dynamic Systems Development Method, or DSDM, arose in the mid-1990s. It was based on RAD and came as a response to the relatively laissez-faire nature of that approach. DSDM aimed to build quality into RAD processes that were forming at the time.

DSDM was originally devised for use in software projects but has since evolved to be a generic approach to project management.

DSDM provides a structured framework of concepts[ix] including a philosophy, principles, processes, roles and responsibilities, products and practices.

The DSDM philosophy is that:

> *best business value emerges when projects are aligned to clear business goals, deliver frequently and involve the collaboration of motivated and empowered people.*

The Signature Dish

Right now, in 2017, the most commonly implemented agile sub-genre is Scrum. This result shows up consistently in several global surveys and papers.

The following is a brief snapshot of just a few of the sources:

- VersionOne's State of Agile report[x].
- ScrumAlliance's State of Scrum report[xi].
- Hobbs & Petit's article in Project Management Journal[xii].

Generally, most business practices become popular because they lead to success and they have a positive impact on the finances of the organisations that use them. Many of the available surveys around project management investigate the reasons why participants choose Scrum and the consensus aligns with this.

Along with the global popularity of Scrum, a new interpretation or meaning is forming around the word agile.

When some people say *agile*, they are often referring to *Scrum*.

This interpretation isn't an entirely correct definition of course. A stickler will gladly point out that agile refers to the overall philosophy. It could also be correctly debated that agile itself is a top-level category or genre of project management.

However, people will do what they want to. Language evolves all the time and we are starting to see this taking place with the interchangeable use of the words agile and Scrum.

***Don't Spook the Herd!* focuses mainly on the application of Scrum.**

If Scrum is the agile method that the majority of people around the world are using, then it makes sense for a practical guide to agile project management to cover Scrum.

Another reason that I have chosen to focus on Scrum is due to the results I have seen and achieved with it. Over my career, 20 years so-far working in and leading projects, I have seen the application of many project management methodologies. Without a shred of doubt, the approach I have observed as being most successful is Scrum.

Some of the projects I have led using Scrum have been nominated for national awards in the categories of project management, innovation and service delivery. A few of them won first prize.

I have used Scrum to run projects with multi-million-dollar budgets where team members were inventing entirely new technologies and applying them in extremely complicated environments. Each time the results were positive.

Most of the projects I have been involved with or managed that used Scrum have been successful. They are still delivering benefits to this day.

Don't get me wrong, Scrum isn't the only solution. Things can and do go wrong. Additionally, not every type of project is a good fit for Scrum nor agile. I cover this further in Chapter 3 and Chapter 23

However, for the sake of concluding the reasoning, let me say this. In the right context Scrum not only works, but it works quite well. Given the sub-title of this book is *How to Get your Agile Projects Running Smoothly*, it makes sense to give you an insight into an approach that is known to work in practice, not just on paper. Additionally, I want to provide you with guidance on agile you can use in any industry, not just in IT/software.

Now that we know about the various flavours of agile, let's consider what type of projects might be a good fit for agile management.

Chapter 3
Agile-Compatible Projects

Many different types of projects in various industries are good candidates for working with agile. It comes down to the characteristics of the project itself.

The following list provides some of the characteristics that contribute to the decision of whether a project could be compatible with using an agile approach:

- project uniqueness;
- requirements uncertainty;
- amount and timing of plans;
- cost of changing direction;
- flexibility of project people;
- participation of key stakeholders;
- leadership comfort with gradual information;
- incremental delivery;
- acceptance of missing out;
- consequence of imperfect solutions;
- project team size.

In this chapter, we go through each of these characteristics to understand how they contribute.

Project Uniqueness

How unique is the project?

Consider a residential housing company that produces a small number of *standard* homes. Each home is designed and built following a template plan. The homes are targeted at the more budget-conscious end of the market. To keep the costs down, the builder limits the amount of variations. There isn't that much difference from one project to the next other than local site conditions and perhaps a re-configuration of a few walls here and there.

These projects probably don't lend themselves to agile. They are not all that novel or unique. This building company regularly completes similar house building projects and hence they really do know what they are doing from start to finish.

Overlaying agile project management in this scenario is likely to introduce an overhead that doesn't really provide much benefit. The iterative plan-output-review style of agile would possibly increase the cost of these projects due to the down-times between iterations. Roughly the same house will still be built in the end regardless.

In a different example, consider a one-off, bespoke, custom-designed, architectural masterpiece home. Let's say the home has been commissioned by a wealthy eccentric individual with very particular and flippant tastes. This project will be unique.

The project owner is likely to want to design things on-the-go and will probably change their mind regularly. Additionally, the owner will want to come in and fastidiously check progress at regular intervals to ensure everything is being produced precisely to their liking.

This style of project is a good candidate to run using agile project management. It really would benefit from the ability to run short iterations, applying the plan-output-review approach.

Each time, our flamboyant owner can step in, check what has been produced, decide what they want next, click their fingers and with a wave of the hand spur the project team onto the next block of work.

Requirements Uncertainty

What level of uncertainty or ambiguity is there in terms of the project requirements?

When the requirements on a project are not very well-understood, that project is a good candidate for agile. The main reason is that the requirements will evolve and change as progress is made on producing the outputs. Once the stakeholders begin to physically see the tangible outputs, they start to get a better understanding of what they really need.

We saw this in the last two examples. The standard-home company ran projects with quite certain, well-understood requirements. The bespoke-home project had requirements that were quite uncertain up-front and likely to change quite often throughout the project.

Let's take another example. Think about an organisation that becomes responsible for a new project in an area it doesn't really have that much experience in. For instance, think about a government organisation, say a department of rural and suburban land development staffed with permanent employees who are all very good at running the normal day-to-day operations.

This group might be excellent at land zoning, environment management, civil engineering, etc. Let's say that this group then gets tasked with running a new project to re-vitalise and renew a central urban precinct in the heart of a city.

This project will be filled with uncertainty for the staff involved. It will likely be their first time interacting, designing and building with inner-city residents, urban ecologists, city planners, commercial business owners, retail outlets, tourism boards and regulatory authorities. There will undoubtedly be several rounds of community consultation, master plans, design concepts, reviews, feedback, evaluation, etc.

The early phases of this type of project would work with agile due to the benefits gained from the incremental plan-output-review approach. Specifically, we're talking about the initial phase(s) of the project that produced the feasibility studies, impact reports, design framework and overall plans that would later be used in the construction phase.

Amount and Timing of Plans

How much up-front planning is necessary for the project outputs? Do those plans need to be static, i.e. unchanged, once they are set?

Projects with Significant Up-Front Static Plans

A project with a significant amount of up front planning that can't change, would probably not be a suitable candidate for agile. In such cases it would be extremely difficult to change the plans down the line once they were agreed and committed to.

Take our urban renewal project example again. There is a significant amount of up-front planning necessary to get to the point where the construction phase of the project can commence.

Once construction starts, it is highly unlikely that much can change in terms of the overall design and plans. Once you start bulldozing the area, it would be very challenging to even consider changing course.

Note the distinction in this construction example. I mentioned previously that the early phases of the urban renewal project, i.e. coming up with the plans and designs, would be suitable to working with agile. However, the actual construction phase would not be suitable to working with agile.

This is an example where you can possibly split a project into separate phases. Or, you could even consider each phase its own separate project. Doing this lets you apply a different project management approach to each phase.

Projects that Don't Require Much Planning at All

Think about a project that did not require significant amount of planning. That is, most of the outputs can be produced with very minimal plans.

In these types of projects, you have fairly low risk, non-complex outputs that could be worked out and produced by the project team as they go.

Using agile on these types of projects could be interpreted by the project team as an unnecessary overhead. Agile would introduce a level of bureaucracy that did nothing much to improve the outputs.

For example, think about a project to run a working-bee at a school. That is, getting members of the school community together to clean up the school grounds and make general repairs.

There would be a lot of work to do but really, it would run just fine without the overhead of agile processes. The work is non-complex, people can figure it out as they go without much risk.

A funny thing about schools is that when people go back to them later in life, some folks have childhood flashbacks.

Depending on how positive or not their school experience was, if you were to start imposing rules and regulations on the team of community volunteers, some of them might begin to resent turning up and you may not get the best out of them.

Projects with Gradual Evolving Plans

Lastly, we consider the type of project where too much up-front planning wouldn't work and could in-fact be detrimental. In this case the project would be a good fit to agile.

This is the type of projects where requirements, designs and details are largely unknowable early on. The information emerges as progress is made on the outputs.

One example where this is the case is experimental or research projects. The direction of such projects and details of the outputs don't become certain until progress is made.

In these cases, it might be better to set some constraints, boundaries and milestone targets up-front. Following that, the team could gradually evolve the plans as the project advances using the iterative plan-output-review structure of agile.

Another example is projects with highly complicated outputs that are difficult to conceptualise and visualise early-on. On these types of projects, project stakeholders only begin to understand their requirements and confirm what they truly need once they start seeing the tangible outputs

Building brand-new software products or developing new information technology systems is a classic example that has these characteristics.

In these projects, you often find that the technologies being developed are unique and involve a significant amount of complicated logic decisions, data processing, security precautions, network communications, etc.

Because these types of projects are so complex, the team members sometimes work best by starting with a general high-level overview of what they are going to build (i.e. the architecture). Following that, they delve into the details of each component closer to when they are about to start working on them.

It is often impossible to know exactly what needs to be produced on one output until the details of other components have been worked out and produced.

You can see in this last example why software projects were one of the first industry segments to really embrace agile.

Cost of Changing Direction

Is there a significant cost to altering course during the project?

Heavy Cost of Change

Continuing our journey to see the various types of projects compatible with agile, we now visit the characteristic of cost of changing direction. That is, what is the impact of altering course during the project.

First, take the example of a project to run a mega-large music festival. In such a project, there is a heavy amount of up-front commitment. One of the main output types in a music festival project is the actual booking / hiring of resources ahead of the planned event date.

There is the financial cost of committing to payments ahead of time to lock-in specific dates for the hiring of:

- venues;
- staging equipment;
- audio-visual and lighting rigs;
- decorations;
- support staff;
- marketing & promotion;
- ticketing;
- vendors;
- etc.

There is also the phenomenal cost of booking big-name artists and musicians to perform. Additionally, there is the reputational cost of ensuring the punters / attendees get what they are promised once the line-up has been announced.

As soon as a music festival starts going ahead you find that changing various aspects comes with a significant cost. Suppliers know that once you are locked-in to a specific date and expecting a certain crowd size, you have very little room to move.

Changing the decisions for equipment and people could result in loss of deposits and possibly even exorbitant prices for alternatives depending on availability.

Changing artists and musicians would also involve a financial cost, but worse, could significantly impact reputation and ticket sales.

It is unlikely that this type of project would be suited to agile since modifying plans after the outputs start being produced would cost too much and could put the financial viability of the project at risk.

Low Cost of Change

Next, consider a project to revise and update the basic design elements and content for an online information resource. For example, let's say the project is to update a website that is run by a local city tourism board to promote their city and provide useful information to visitors.

Assuming the project has decided to keep the underlying technology of the website unchanged, then the major outputs of this project are the updated content (i.e. the words and images) and updated design elements (i.e. colours and graphics).

In this project, there is quite a low incremental cost of change. There is little amount of financial expenditure or reputational investment that is *sunk* or *fixed*. Most of the expenditure is *variable*, associated with what items the project team is going to work on next.

Essentially, if the owners draft an update or an idea and then try it out, they can quite easily modify certain aspects as the project progresses. Additionally, given the information is published electronically online, it is relatively cheap to change.

It would be quite feasible to run this project using agile and break it up into distinct work blocks that focussed on delivering various updates each time.

Flexibility of Project People

How flexible are the main project people in terms of adjusting targets and designs as the project proceeds?

This characteristic is essentially a no-brainer (i.e. easy to understand). If the main people on the project are flexible when it comes to adjusting targets and designs as the project proceeds, then you have a good candidate for agile.

When we say project people here, we are referring to the project team and the main group responsible for setting the direction on the project.

If the project people are inflexible, below is a hypothetical example of what could happen if you were to attempt running agile with them.

First, it would be very difficult to gain their confidence early on since they wouldn't be able to see the detailed plans up-front about what was going to be produced, when, how long it would take etc. They essentially wouldn't want to believe in the approach.

Next, when you started to attempt getting into some of the agile mechanisms with them you would find it difficult due to their scepticism.

Following that, once the project started progressing you would see the inevitable design revisions and direction changes that are perfectly natural in agile projects. Once this started occurring, your project people would begin to resist the changes.

And the pattern would continue. The result would be a group of people with less-than-ideal commitment and motivation toward the project. Not a very good combination if success and collaboration are things you are aiming for.

Participation of Key Stakeholders

Put a microscope on the name agile and ask why that name was chosen. The reason is to show the focus of a style of managing projects that allows adjustments and changes to produce outputs more closely aligned with what the key stakeholders need.

The project is agile, i.e. adjusts, around what gives the most value to the key stakeholders.

Now, for this to work, the project needs stakeholders to fully participate.

Leadership Comfort with Gradual Information

How comfortable are the leaders of the project with gradually evolving levels of information and progress?

On a similar line of thinking as the previous sections, this characteristic again focuses on the feelings of some of the people on the project. This time it is the project leaders. That includes the person(s) that provide the funds for the project, the person responsible for owning the project and making major decisions, along with the person(s) that represent major groups for the project.

Ultimately this is the group that sets the tone of the project. If this group is supportive, then that support usually cascades through the rest of the people on the project.

If the project manager has a good relationship with this group and holds their confidence, things will go well and support will be there when it is needed.

If the leadership group on a project is uncomfortable with emerging information and progress, then it is unlikely that agile project management would work with them.

By its very nature, there is little information up-front in an agile project and minimal documentation overall. The information emerges when the project progresses, outputs are produced and stakeholders begin providing feedback.

If this doesn't sit well with the project leaders, you are at risk from the get-go.

That is, they won't really have any confidence in the process from the start and thus as soon as issues start arising on the project, you might find yourself in a situation where the leaders play the "I told you so" game.

Worse, you could find they are unwilling to assist since they just don't want to be associated with the project.

Acceptance of Missing Out

Are the stakeholders OK with the idea that, most-likely, the project will not end up producing every single output they desire?

Have you ever heard the phrase "one's eyes are bigger than one's stomach"?

When people involved in a project start thinking about requirements, designs and solutions, they often come up with more ideas than a project can deliver.

On an agile project working with fixed constraints, it is likely that the list of desired outputs in the backlog is larger than what is physically possible to create. Some outputs just won't be completed.

Depending on how the stakeholders feel about this contributes to whether agile will be a good fit for them.

Agile was designed in part to respond to the issue of stakeholders wanting more than they could get. Agile deals with this using the concept of prioritisation.

The set of desired outputs is put into a backlog and ordered by priority. The most desired output is at the top of the list and the least desired output at the bottom. The project team then works through the list from top to bottom.

The project team will produce as much as they can within the constraints of time, budget and team capacity.

Inevitably, the team won't make it through to some items on the list. That is a side-effect of constrained resources. In an ideal world, or should I say fanciful, stakeholders should be able to get everything they so desire. In reality, that just isn't possible and people need to make trade-offs.

Agile projects explicitly acknowledge this trade-off up-front. No promises are made as to exactly what will be completed by the end of the project.

Instead, the project commits to producing as much as the team has capacity for. The project gives preference to delivering outputs in priority order and responding, as necessary, to the evolving understanding of the stakeholders.

Project Team Size

Agile projects tend to do better with relatively smaller team sizes. The issues in this context are communication and collaboration.

In agile projects, there is a need for the team to regularly be in close proximity with each other. Whether it is physical, or virtual, agile teams need to regularly communicate with each other and collaborate on the production of outputs.

Agile projects have a framework of principles, work practices and activities. The larger the project team, the more challenging it can be for these collaborative approaches to work effectively.

For example, it is easy for a relatively small team to have a daily get-together where they share their progress and discuss issues. As the team size grows, however, these simple little catch-ups can end up becoming unwieldy.

There is a nice formula that we often refer to in project management regarding lines of communication. N(N-1)/2 is the number of communication channels on a project where N is the number of team members.

Each time another person joins the group, the number of lines of communication between everyone else on the project goes up exponentially.

For two people, you have one line of communication between each other. For three people, you have three lines of communication. Four people, six lines. Five people, ten lines, and so on.

What's the magic number? I have run agile projects with teams of around 20 people quite successfully. I have also done so with tiny projects with just a few people.

As the number of people grows you need to consider the practicality of all members of the same team communicating and collaborating with each other.

Incremental Delivery

Is it acceptable for the outputs of the project to be delivered gradually in increments?

Certain types of outputs can be gradually evolved through a series of stages. One way of thinking about solutions like this is as prototypes.

The first few attempts at the solution are known as drafts. They are designed to try out various ideas and concepts and give the end users something tangible to look at, play with and think about further.

These types of outputs progress through a series of stages. They go from that original draft or prototype to the last version that they agree can be called *ready*.

Solutions where it is OK to deliver in this way are well-suited to agile projects.

This approach assists with the users gaining a tangible and deeper understanding of the final outputs in a progressive manner.

Each time the output is produced or updated, the user can build on their knowledge and understanding of it. This helps provide feedback for further elaboration of the requirements.

One type of output like this is a document or publication. For example, this book you are reading has been produced incrementally. Each draft I created was taken through an editing process as well as reviews with various sample audiences. Eventually, the editors, reviewers and I settled on the content that we were happy to release as the first edition.

Another type of output that lends itself to this approach is software. This is the reason agile was initially heavily associated with that industry.

Due to its digital nature, software is difficult to visualise initially but easily updatable. Software can have many features, sometimes representing highly complex data and information structures. Therefore, it can be hard for people to create a mental model of what a finished product might look like and how it might work.

In response to this visualisation problem, the industry has found it better, in some circumstances, to create working versions of outputs and put them in the hands of end users early.

As the users become familiar with the tangible outputs they then provide feedback on further improvements that could meet their needs and provide additional value.

Consequence of Imperfect Solutions

How critical is it that the outputs of the project work perfectly the first time?

This characteristic is closely related to the previous regarding incremental delivery and amount of up-front planning needed.

The idea we are focussing on here is known as the criticality of the project. In other words, what is the consequence of the solutions of the project not being perfect the first time?

There are circumstances where it may be necessary to deliver the outputs of a project in one single go.

These are the types of project where you usually only get one go at the end-result. That is, the final outputs that are delivered need to be correct the first time. Other phrases to describe these types of projects are *big-bang*, *main-event* or *moon-shot*.

These types of projects don't really align with agile project management. The requirement to be precise the first time means the gradual incremental delivery of agile isn't feasible.

A type of project with high criticality is one with outputs that could endanger life if something went wrong.

For example, the creation of some type of medical intensive care emergency support device. The project is probably going to want as much of the output thought through and planned in detail before the device is built. It can't really be created as a rough version and rolled out for use in a hospital setting if it is only partially ready. The risk is too high that someone could be seriously injured or die as a result.

Another type is significant, high-use, engineering structures.

For example, you can't really build a new bridge over a deep valley incrementally. It must be perfect the first time and then last as long as possible. The safety of people travelling over the bridge is paramount. Having the project team try out various designs and materials for, say, the footings, pillars and side-railings on the finished product, and then asking for feedback, just wouldn't be acceptable.

Another type of project that fits this category, as the name mentioned earlier suggests, is one which sends people into hostile environments. Flying people to the moon in a rocket is an example. There is no room for error. Every single output of the mission needs to be correct the first time so that the people can come home safely.

We're all Matchmakers

There you have it. We now have a good idea about some of the characteristics of projects that make them suitable, or not, to agile project management.

Next time you find yourself involved in the early stages of setting up a project, have a think about some of the descriptions and examples we went through here.

If you see a good match across many characteristics of your own project, then perhaps it might be a good candidate to run using agile project management.

On the flip-side, if you find that agile project management is going to be used, but you don't see a good match, it would be a good idea to speak up and share your thoughts with the project leaders.

This concludes Part I. We should now have a shared understanding of what the word agile means. We're ready to go-forward together.

In the next part, I open the door to aspects of agile projects that often get thought about late in the process. It's time we acknowledged the real driver of agile project success.

PART II
PEOPLE

Chapter 4
Project People

Until we see amazing leaps in artificial intelligence and autonomous universal replicator technology, projects are going to be carried out by people, for people. Agile projects are no different.

As much as we'd all love for machines and software algorithms to do the work of planning and running projects for us, it just isn't going to happen in the foreseeable future.

Similarly, we don't just run projects for their own sake, we run them to deliver outputs for people. Without going too far into it, the high-level lifecycle for any successful project goes roughly something like this:

- A person or group of people have some need they want fulfilled or problem they want solved.
- Someone with enough money decides they will take up the challenge and run a project.
- As the project progresses, people produce outputs.
- Once complete, those outputs are utilized by people.
- The utilization of outputs leads to outcomes.
- The outcomes eventually result in benefits for … you guessed it, people.

This description provides us a critical point: People are the key to successful projects.

It is a point that you will need to keep coming back to, so I won't mince my words. Say it again with me one more time: *People* are the key to successful projects.

Things like principles, strategies, processes, tools, technologies, techniques and tactics are important enablers for a project. However, if all is not well with the people, none of those other things matter (see acknowledgement 2 regarding this, at the back of the book).

The first step in your journey toward agile utopia is to think about the people you will be working with. This begins with the person or people funding the project, ends with the people receiving the overall benefits and includes everyone else in between.

In this chapter, we look at the various roles or groups of people involved in agile projects. It is important to consider each of these groups since they will each have their own part to play. They will also have their own separate needs when it comes to understanding and becoming comfortable with the agile process.

If you have been around projects long enough, you will notice that many of the roles described in the following sections are not unique to agile and could easily be applied to all types of projects in general (see acknowledgement 1 as well as Zwikael & Smyrk[xvii]). There are many aspects of agile projects that follow more general project management principles.

Funder

Also known as the Sponsor, the Funder is the person that provide the resources and initial drive to start the project. This is the one that makes the project possible by providing the finances, i.e. the person with the money. The funder is usually at arms-length from the day-to-day details of the project. They are involved but more at a high-level point of view. They will want to know their money is being spent wisely, efficiently and will deliver what it is intended to. The funder will also want to know of any major risks and issues associated with the project.

Owner

Also known as the Project Executive, the Owner is the individual that has the day-to-day decision-making authority on the project. Why? Because they are also responsible for ensuring the project ultimately achieves its target outcomes and provides the associated benefits. Sometimes you will see the same person undertaking the role of funder and owner, but this doesn't always have to be the case.

Picture someone in a C-Level role, i.e. Chief, in a large organisation. Your project will be just one of many responsibilities and strategic activities that person is leading. In cases like that, they will most likely hand off the day-to-day responsibility for owning the project to a separate person, i.e. the project owner.

Project Manager

This is the role that a large proportion of people think of when talking about projects. The Project Manager is the person that will be responsible for planning, coordinating, monitoring, controlling and closing the project. Think of the project manager as the *central hub* that helps connect and guide the other roles on a project.

A well-respected leader in the field of management research whom I had the privilege to learn from and work with, Dr Walter Fernandez, once described the role of project manager as one-part orchestra conductor and one-part lion tamer. After a few years' experience, I realised exactly what he meant.

In a successful, well-functioning agile project, the owner and the project manager will have an effective, collaborative and collegiate working relationship. Together, these two roles are the key enablers for the project.

Just like a couple in pairs figure (ice) skating, the project owner and the project manager need to work well together and trust each other for a smooth performance.

Board

A Project Board is sometimes also known as a Steering Committee, Governance or Leadership Team. This is the group of people that provide important input, opinion and guidance to the management of a project.

Don't be put off by the formality of the name. Depending on the nature of your own project and organisation, this group can be as formal or informal as you want.

They could meet around a round table if it suits, or, they could just as easily catch up together in a local coffee shop. Honestly, go with what works best for your organisation, their values and expectations.

The main goal is to ensure that there are specific people nominated to make important decisions on the direction of your project.

The group is usually made up of a small number of leaders that represent the views and responsibilities of key stakeholders for the project.

Having a well-functioning board is a great way to ensure your project is delivering the right things for the right people in the right way. A good board provides a level of transparency, consultation and collaboration which ultimately serves a project well.

Team

This group consists of the people directly assigned to create the outputs of the project. A project team can be made up of various roles with any number of skills depending on the type of project and on the organisation running the project.

The team members will have different levels of time-commitment which the project manager will need to be aware of. You could see teams of staff who work on the project for the entire duration, teams made up of various people that come and go on the project depending on what they work on, or some combination of both.

Something to consider here is including a sufficient number of experienced professionals in the project team.

By its very nature, a project is a one-off unique activity producing outputs that, more than likely, won't have been produced in the same way before. As such, you'll want at least a few people that are able to self-organise, innovate with limited information and guide others they work with.

Customers

Also known as Users, this group is made up of people that will utilise the outputs produced by the project. As we eluded to earlier, it is when customers start utilising the outputs that we begin to see the target outcomes of the project being achieved.

Be aware of the time delay between project completion and the customers utilising the outputs. The normal scenario is that customers don't usually get to start utilising outputs in real day-to-day operations until the project is complete.

Why is this important? Because of this quirk of time, someone other than the project manager needs to take responsibility for the long-term achievement of target outcomes for a project. It is recommended that the project owner takes on this responsibility.

Close attention needs to be paid to the customers early in the project and as outputs are being designed and created. The project needs to maximise the value customers get from the outputs so that they naturally want to utilise those outputs.

One of the drivers behind agile project management is ensuring customers get value from the outputs.

If customers naturally want to utilise the outputs, the project is more likely to achieve its target outcomes and deliver benefits in the long run.

Beneficiaries

This is the individual or group of people that benefit in some way from the achievement of a project's target outcomes. It is a very broad group and could include various people or roles depending on the nature of each project.

Like the case for customers and target outcomes, there will normally be a time delay between the project completing and the generation of benefits.

Examples of possible Beneficiaries include:

- Investors for private / commercial projects, e.g. an entrepreneur benefiting from the financial return on a project they funded to develop a new product or service.
- Demographic groups for large scale national &/or international projects, e.g. families in third-world conditions benefiting from a project to develop technology that simplified purification of local drinking water.
- Specific communities for projects that target particular problems, e.g. farmers of a crop benefiting from a project to develop new methods of pest control for that crop.
- Members of the public for government sponsored projects, e.g. the entire population of a country benefiting from a national project to reduce significant amounts of revenue lost through multinational profit off-shoring and tax avoidance.

- Customers of organisations and internal staff in projects to improve tools or work practices within an organisation, e.g. external customers and internal first-level support staff benefiting from a project to improve support request tools used in a company or government department.

Suppliers

This group represents anyone that is not directly part of the organisation's day to day project team but who provides outsourced/procured products or services to the project. That is, the external vendors for the project.

Examples of suppliers include:

- providers of materials &/or equipment used to create the project's outputs;
- specialised sub-contractors or consultants that participate briefly to provide their expertise in certain areas.

Impactees

Last but not least, Impactees are another broad group of people that are associated with the project. They don't work on the project and they don't explicitly gain benefits from the target outcomes. In other words, impactees don't fit into any of the other categories. This group may be affected either positively or negatively. That is, they may gain a fortuitous windfall, or the project may be detrimental to them in some way.

It is important that the you work toward empathising with the feelings of impactees, especially those affected negatively.

Techniques that can be utilised in support of this include consultation sessions, change management activities, forums for voices to be heard, places for thoughts to be expressed and mechanisms for ideas to be shared.

Implicit in some of the agile project management practices are opportunities for review and feedback. This should always include working with impactees.

As you start bringing these groups of project people together, a culture unfolds. Specifically, it is the way they relate to each other, to projects and to project management practices.

In the next chapter, we take a detailed look at this culture and why it is important in relation to agile project management.

Chapter 5
Check the Culture

In this chapter, we look at how the culture of people impacts the success of projects. We also look at how culture and agile project management relate to each other and why it is important to consider.

Let's start with a definition. The Oxford Dictionary defines culture as follows:

the beliefs and attitudes about something that people in a particular group or organization share

Our attention here in this chapter is on the beliefs and attitudes of project people in relation to:

- prioritising;
- dealing with uncertainty;
- focusing on outputs;
- documentation;
- managing risks;
- collaborating;
- problem solving;
- delivering completed work;
- self-organising;
- reviewing progress;
- responding to feedback;
- evolving plans;
- sharing information;
- supporting each other;
- reflecting;
- learning and improving.

You would have to agree there is quite a lot to culture when you see a list like this wouldn't you? But let us not shy away from it. The reason we dedicate an entire chapter to culture in this book is that it can make a big difference to the success of agile projects.

Culture Through the Lens of the Agile Manifesto

Recall back in an earlier chapter where we introduced the Agile Manifesto. Along with the manifesto itself, there are a set of twelve principles that begin with the words *we follow these*. This is a good example of agile-supportive culture in action. It is the reason for using these principles as a reference point in this chapter.

Once again, from the original descriptions, let's replace the word *software* with the word *outputs*. That way we can apply these principles from a more general point of view to any type of project.

Prioritise Customer Satisfaction

Our highest priority is to satisfy the customer through early and continuous delivery of valuable [outputs].

Agile projects work well when there is a culture of making customer satisfaction the main goal.

Consider an organisation where the culture is for sales or profit to be the primary goal. In such an organisation, it is possible that projects are run where customers are relatively less important and where the project team is quite ruthless in its drive for sales.

The result might be outputs that sell well initially and provide a good financial return, but this won't last. If there isn't a good match to customer value, then you risk not achieving target outcomes and ultimately a reduction or loss of benefits.

Alternatively, if the project is focussed on customer satisfaction and the value they obtain from the project's outputs then there will be a lasting connection and thus better results in the long run.

Welcome Changing Requirements

Welcome changing requirements, even late in development. Agile processes harness change for the customer's competitive advantage.

The culture here is one of welcoming change. Seeing change as a positive contributor to a project's success.

A culture where the project people are uncomfortable with changing direction is unlikely to be a good fit with agile project management. By its very nature, an agile project is all about changing direction based on customer feedback.

The reason why this culture of welcoming change works with agile is that the people react positively when feedback comes in. Even if that feedback would alter the plans. People see it as an opportunity to produce outputs that are a better fit to customer needs.

Deliver Working Outputs Frequently

Deliver working [outputs] frequently, from a couple of weeks to a couple of months, with a preference to the shorter timescale.

A common challenge in projects is getting the team to finish what they are working on and release it out to the world for customers to start using.

One of the underlying reasons for this issue is that you often employ highly experienced professionals on projects and these types of people are often quite meticulous in their work. It is this attention to detail that make them good at what they do but it can also be their Achilles' heel.

You occasionally hear statements along the lines of "it is not good enough yet to release" or "there are still a few issues we need to resolve".

The culture in this example scenario is one where the project team members believe they know exactly what the customers and stakeholders need.

Agile projects on the other hand put the knowledge of what is best back into the hands of the customer.

A supporting culture for agile in this context is one where the project team regularly finishes whatever they can in a given timeframe. They aim to deliver the working outputs to the customer as soon as possible.

A key element in this culture is the definition of *done*. The output handed over needs to be usable by the customer. The culture of the project people needs to be such that they are comfortable focusing on regularly delivering completed work.

Endless tinkering, perfecting and gold-plating are discouraged. Instead, encouragement should be on getting the outputs into the hands of the customers.

The idea is that customers tell the project people sooner rather than later whether the outputs meet their requirements and provide value.

Work Together Daily

Business people and developers must work together daily throughout the project.

In this principle, the term *business people* refers to the folks on the project that are not physically producing outputs. They are the ones providing the information and details other people will use to create the outputs.

Those other people, i.e. *developers*, are the ones that physically do the work to create the outputs.

The *information and details* include descriptions about what customers want to use the outputs for and what they would like to achieve by doing so. It also includes the constraints and dependencies that might exist as well as any rules that must be adhered to.

The culture that is encouraged in this principle is that of regular collaboration between people setting requirements and people creating outputs to meet those requirements.

That first principle of agile we discussed above, *prioritise customer satisfaction*, is greatly supported by this principle.

An organisation where, daily, the people defining the requirements are working together with the people creating the outputs, is one that is more likely to produce a good match for what customers and stakeholders need.

Think about the opposite situation, one you may have come across already in your own professional career.

Let's say a group of people are looking for certain outputs to be created. Let's also say that this group finds the right people to create their outputs for them but all they do is tell them roughly what they want toward the start of the project.

They leave the people producing the outputs on their own. After all, those *workers* should just be able to figure it out on their own, right? That's what they're paid for, right?

Can you see where this last scenario is going? It is an example of an organisation with a culture where the people defining the requirements of the outputs don't really collaborate with the people creating the outputs.

Flitting in at the start of a project, sprinkling statements about what you want like a prima donna and expecting things to turn out correctly just like that is a fairly naïve management style.

This is, most likely, one of the sources behind countless project failures all over the world. It is this type of misled belief and attitude to creating outputs that agile aims to resolve.

Support and Trust Motivated People

Build projects around motivated individuals. Give them the environment and support they need, and trust them to get the job done.

There are several culture elements in this principle. The first centres around how people are chosen to work on projects. The second is the culture of enablement. The third focuses on the culture of trust.

Let's look at the first: how people get selected to work on a project. In some organisations and indeed some geographical regions, staff are sometimes selected to work on projects for reasons such as:

- rank / hierarchy;
- gender;
- nationality;
- social status;
- age;
- tenure / years of experience;
- friendship / connection;
- ability / non-ability.

I hesitate to comment on the underlying merit or otherwise of these reasons other than to say there is another way.

Effective agile projects suggest a culture of staff selection based on motivation.

That is, give preference to the following characteristics when it comes to selecting staff to join project teams:

- energy;
- enthusiasm;
- positivity;
- drive;
- ability to complete work.

This doesn't suggest you overlook other important characteristics such as skills and experience; these attributes are also very important.

It merely suggests that organisations with a culture of motivated staff are the ones more closely aligned to working well with agile project management.

Next, let's consider the second element of this principle: organisational culture around support and enablement.

What better way to feed into the energy and enthusiasm of motivated project people than to provide them with an ideal working environment and support them throughout the project.

The concept of environment and support here can mean many things.

There are the tools and processes the team needs to be productive. This includes the tools for planning and tracking their work, the tools for managing the knowledge and information of the project, and the tools and processes for producing the outputs. That last group could be anything and is usually quite specific to the type of project and industry.

There is the physical space where the team works together daily, i.e. the desks, meeting rooms, offices. There are also the technologies to support collaboration between remote / physically-separate teams. This includes PCs, internet connections, chat applications, telephones, video/web cameras, display screens, web conferencing technologies, etc.

Lastly there is the environment of focus. Enabling the team members to focus on their project free of distractions. This includes not being allocated to numerous projects at once, being relieved of normal day-to-day operational duties, and not being overloaded with too much work and unrealistic expectations.

When it comes to support, the culture that works well with agile project management is one of nurture and care.

Project team members should feel comfortable asking for help and they should be confident they can rely on assistance when needed.

The third element in this principle is the culture of trust. This one aligns closely with the previous aspect of support. The culture of trust goes both-ways. That is, the project team members should be able to trust that managers, leaders and other stakeholders will be there for them when they are needed. In return, the managers, leaders and other stakeholders should be able to trust that the project team members will deliver results and get the project completed.

Notice I used the words *should be able to trust* in the paragraph above.

For managers, leaders, stakeholders and project teams to trust each other, they need to display behaviours that enable and reinforce that trust.

Agile project management works best when there is a culture of enabling and valuing trust among all project people.

They say trust is a two-way street and this is true in agile projects.

Speak to Each Other, In Real-Time

The most efficient and effective method of conveying information to and within a development team is face-to-face conversation.

While you may think it obvious that people on projects talk to each other face-to-face, this principle exists to cater for the fact that it doesn't happen effectively in many scenarios.

Let's go through a few examples of situations where you see the opposite culture of people on projects not speaking to each other enough.

The first example tends to show up on projects in industries and sometimes countries or regions where there is also a culture of litigation. That is, where people and organisations tend to take each other to court. Because of this fear of being involved in legal proceedings, people on projects tend to communicate in written form, i.e. they put it in writing, so it can be pointed back to later as a defence or claim. What you observe in this situation is people tending to avoid communication through conversations.

The next example appears in bureaucratic organisations where there is a belief that projects should create a detailed record of all activities and decisions. For various reasons, projects in this type of environment tend to include people that prefer everything to be written down in documents. What you end up with is project team members spending the bulk of their time typing up requirements, plans, designs, specifications, architectures, decisions, requests, reports, etc.

The third example is where there is an over-use or over-reliance of communication technologies. This happens when project teams spend most of their time communicating to each other via email, chat, sms, messaging, forums or other similar tools. In these type of projects, there seems to be an aversion to people physically talking to each other, using their voice and ears. People tend to avoid activities such as walking up to someone else and asking them a question. The impact is misunderstanding and delays between communication. The reason is the asynchronous, i.e. not synchronised, nature of the technologies in use.

To counter this, a culture where projects encourage real-time conversations between people is one that would support agile management. The benefits you get from such human interaction include:

- Faster conveying of messages: people say something and it is heard instantly.
- More likelihood of accurate understanding of meaning: people can check their understanding of something immediately using active listening questions, i.e. "do you mean ____?".
- Picking up on body language: seeing how people feel about a conversation through the visual cues of their facial expressions and body movements.
- Relationship building: people tend to form stronger bonds when they are in each other's presence.
- Shorter turn-around time for adjustments and changes: people can talk together about solution ideas rather than typing them into documented form.

That last point is one of the keys to good communication with agile. A culture of collaboration via conversation makes it much easier to review results and adjust plans as a project progresses.

Focus on delivering working outputs

Working [outputs are] the primary measure of progress.

Nothing trumps outputs when it comes to progress on projects. The focus here is on that key word at the front: working. It is possible, and sometimes necessary depending on the nature of the job, for project team members to spend a long amount of time ensuring that the outputs of the project are *just right.*

Agile projects on the other hand take the approach that project team members are not always the best interpreters of what *right* actually is. Instead, the project's customers and stakeholders are the ones best-placed to make the decision on suitability and fit.

A culture that focuses on delivering working outputs for customers and stakeholders to evaluate is one that fits with agile.

Included in this principle is the question of progress. What is it that tells you a project is moving along its schedule toward its goal?

In some organisations, you see a culture focussed on calendar dates and arbitrary milestones as progress measures.

This is all well and good but merely handing over a set of outputs by a certain date or ticking off a certain milestone isn't a good measure of progress.

If the outputs don't work or the milestones don't have much to do with customers end-use of outputs, then the feedback will be minimal and of limited benefit.

Instead, a culture that focuses on completing working outputs that can be utilised by customers is one that supports real progress. Such a culture encourages project team members to zero-in and deliver what the customers and stakeholders really want.

Team members spend less time on activities that provide less advancement toward the desired result of customers utilising outputs, so they can generate target outcomes and benefits.

Promote sustainable work practices

> *Agile processes promote sustainable development. The [key stakeholders], [project team], and [customers] should be able to maintain a constant pace indefinitely.*

If you have ever heard the phrase "just get it done, no matter what the cost" on a project you probably understand that this usually results in something detrimental to the project.

A culture where project people think working overtime is OK and just something that happens now and again fails to see the impact that overtime has on people.

Sure, overtime might be lucrative for team members occasionally but over the long term working overtime can impact the following:

- energy levels;

- motivation;
- family and personal relationships;
- physical health;
- mental well-being;
- cognitive ability;
- etc.

Additionally, a culture where the project people are not careful with the resources available to them, i.e. money, time, materials, etc., is wasteful and may end up producing less-than optimal results.

This principle suggests that agile projects work best when there is a culture of respect for people and for the other resources available to a project.

In such a culture, there is acknowledgement of the following:

- There is a finite amount of people on the project.
- The people have a given set of skills and capability, i.e. they can only deliver as much as they are able to.
- There is, usually, a finite budget and timeframe within which to complete the project.

When the people on the project are OK with this, then you have a good fit with the various agile processes. The main ones being delivery of outputs in priority order, working in a series of fixed time intervals and not attempting to do more than physically and logically possible.

Get this right, and it is indeed possible for the project people to continue working together at a steady pace.

You are more likely to see the project people converge toward that sense of *flow* where they are immersed in a feeling of energised focus, full involvement, and enjoyment in the project.

Stay Professional

Continuous attention to technical excellence and good design enhances agility.

The culture discussed in this principle is that of professionalism. True masters in a craft, trade or specialty are ones that are meticulous when it comes to performing their work. Such professionals have attributes and work practices where they:

- keep their technical and non-technical skills up-to-date;
- utilise up-to-date tools and materials;
- assess and refine their own processes;
- seek feedback and adjust accordingly;
- design solutions that are a good fit for actual requirements;
- consider future scenarios in their designs to allow for future enhancements or upgrades.

With such professionals, there is a culture of continuous improvement in everything they do. It is this culture that aligns well with agile project management.

Consider the opposite, a case where the culture is something more akin to "near enough is good enough" or "we know best". It isn't difficult to see that this wouldn't really work with the agile plan-output-review cycle.

In this case you would end up with poor results each time you reached the review stage. There would be a less than optimal fit between what the customers and stakeholders required and how the project team delivered on those requirements.

Do only that which is necessary and sufficient

Simplicity--the art of maximizing the amount of work not done--is essential.

This principle closely relates to the question of "how much is enough?". An organisation where the culture is careful with the use of its limited resources is one which can maximise the outputs possible from those resources.

You sometimes see project team members getting bogged down in unnecessary activities when producing outputs on projects. Examples include excessive meetings, marathon design sessions, too much documentation and over-the-top status reporting.

In these cases, the main project resources of time and money are spent in ways that, while perhaps useful, are not necessary or sufficient to deliver actual working outputs.

Instead, when the project people are careful with what they work on then you will see a higher likelihood that the project will get more bang for its buck so-to-speak.

If the project team only works on the things that are necessary to meet the needs of stakeholders, they are on the right track.

At the same time, if the team ensure that what they work on is sufficient for the project's stakeholders, then they will minimise the amount of wasted or unnecessary effort.

Don't Micro-manage

The best architectures, requirements, and designs [i.e. solutions] emerge from self-organising teams.

We will touch on this topic a few times in the book. To summarise it here. Project teams generally produce their best work if they are trusted and left to get on with the job.

Trust in this context includes belief that project teams can come up with appropriate solutions. Trust also includes autonomy for the team to work together, without continuous interruption or meddling of managers and other project people.

A culture of self-organisation and trust is a good fit with agile project management practices. In return, there are various mechanisms built-into agile practices that help ensure that trust is deserved and maintained. One example is the regular plan-produce-review cycle. These regular check-ins with the project team to see the progress of outputs encourages team members to deliver. It also picks up any issues early should they occur.

Embrace Self-Reflection

At regular intervals, the team reflects on how to become more effective, then tunes and adjusts its behaviour accordingly.

Closely related to the previous section on technical expertise, this principle covers the culture of self-review and honesty.

Project teams that regularly review their own abilities and effectiveness are ones that improve and deliver better results over time. When project team members can take a good honest look at themselves and how they are performing on the project, they can self-detect issues.

They say awareness is the first step toward change. In this context, the *change* is improved project results and *awareness* is of the issues holding the project team back from achieving the best they can.

Clearly, a culture of self-reflection and continuous improvement is something to strive for in any professional activity.

Does Your Culture Fit?

Have a think about culture next time you are considering using agile project management or if you happen to find yourself on a project that is already using it.

Think about some of the various cultures we have spoken about in this chapter and ask yourself: "Is there a good fit between the culture of this organisation and agile project management?".

If you do think there is a good fit, congratulations. You're one step closer to getting your agile project running smoother.

If you don't think there is a good fit, ask yourself what you would like to do about it. Perhaps it is unlikely that you will be able to influence the culture yourself in which case you might like to reconsider using agile project management for now. Or, maybe there is a way that you can speak to the various leaders in your project and together you can try to nudge the culture toward being a better fit with agile project management.

Speaking of leaders, in the next chapter we look at why it is wise to consider the feelings of project leaders. We investigate the ways that leaders' feelings toward agile can make or break your project.

Chapter 6
Look After the Leaders

A very important contributor to the success of an agile project is how the leaders *feel* about it.

Aspects that contribute to a leader's feeling toward an agile project include:

- their opinion about projects and project management;
- previous experience with agile;
- the scale and importance of the project;
- relationships with other stakeholders;
- level of risk for the project;
- their relationship with the project manager;
- transparency of status reporting;
- adherence to schedules;
- sensitivity to modification of plans and priorities;
- degree of comfort releasing partial sets of outputs.

In this chapter, we consider these aspects, discuss why they are important and provide a few suggestions for how you can help.

While the topic of this book is agile, much of what we cover in this chapter could be applied in a general sense to any form of project management.

Who Are These Leaders?

We touched on this earlier in Chapter 4. There, we included:

- the person that provides the funds for the project;
- the person responsible for owning the project and making major decisions;
- the people that represent major groups for the project.

Now that we are getting down and deep into the topic of leaders here in this chapter, let's list exactly who we are talking about. Project leaders include the following:

- funder;
- owner;
- project board members.

What's So Special About Them?

In Chapter 3 we mentioned that leaders set the tone for the project. Their support for the project, or otherwise, usually cascades through to the rest of the project people. This is the essence of it. If the leaders feel positive about the project, then the likelihood of success increases.

A leadership group that supports the project is more likely to:

- ensure the project has the resources it needs;
- assist with treating project risks;
- help resolve issues when they arise;
- spread the word about how important the project is;

- encourage other people to contribute to the project;
- highlight project successes;
- promote utilisation of outputs so there is a greater chance of achieving the project's target outcomes and receiving the subsequent benefits.

What Contributes to Their Feelings About the Project?

Opinion About Projects and Project Management

Have you ever had a conversation with a friend or colleague about something you really believe in and then found out the other person is quite sceptical about that thing? How did the conversation go?

Let's use a trivial example to get to the heart of why this opinion matters. Imagine yourself walking up to your friend one day and saying "hey, I've got a spare ticket to the big event on the weekend, would you like to come?".

Assume the *big event* is something like a ballet, or a basketball game, or movie launch, or theatre production, or whatever. Assume also that your friend has a bunch of pre-conceived notions about the big event, perhaps negative or uncertain, but they accept anyway.

Your friend is likely to be a bit stand-offish about going along with you and the experience is probably going to be a bit tense early on, right?

You never know, they might end up hating the big event or they might turn out to really enjoy it. It depends on how open they are to try it out and how good the performance of the big event turns out to be.

Now, apply a similar example to a scenario about a leader and their opinion toward projects and project management.

Take a case where one of your leaders has a bunch of pre-conceived opinions that are once again negative or uncertain. What sort of feelings do you think would be going through their mind regarding the project?

- Are they going to want to be involved?
- Will they be a bit stand-offish early on?
- Is their level of enthusiasm going to be noticed, and possibly reflected, by others around them?

Consider this next time you are involved in a project that is in its early stages. See if you can get a sense for the leaders' opinions about projects and project management.

Or, if you are a leader yourself, have a think about whether you have an inherent bias within you.

In either case, try and visualise how the project might pan out as a result.

You never know, just like the big event, the leader could end up disliking the project as they expected. They could also turn out to absolutely love the project and become an ardent supporter!

Later in this chapter we will talk about some of the things that might help in nudging their feelings toward being supportive.

Previous Experience with Agile

OK, now that we're into the swing of this visualisation technique, let's apply the same thinking to another important aspect: a leader's previous experience with agile project management.

- Has the leader previously been involved with an agile project?
- How did the project turn out? Was it successful or was it a failure?
- Was it a one-off case, or have they been involved with many with the same end-result?

You can see where we are going with this example right?

First, a leader that has had no previous experience with agile project management is probably going to be a bit hard to read.

It depends on their own personality traits and how they feel about trying new management practices.

You could be somewhat certain that, given they are in a leadership position, they would have enough experience to have seen a few other business and project management practices come and go.

Their opinion about this next one, agile, is likely to be swayed by how successful or otherwise their previous experiences were.

In a similar line of thinking, consider a leader that has had previous experience with agile project management.

In this case, they are likely to come to the party with some inherent bias toward this next agile project based on how successful or otherwise those previous agile projects were.

Scale and Importance of the Project

We're on a roll now. This time we consider how big and exciting a project is. For a busy leader, with many responsibilities and career aspirations of their own how do you think they are going to feel about a project based on its scale and importance?

- Is this a big project or is it small?
- Is the project very important to the organisation or is it relatively unimportant?
- Does the project have a big budget, or will it be limited by a small amount of available funds?
- If they work with this project, will it help contribute to the leader's career in some way?
- Will it be a high-profile project within the industry and possibly even in the public domain?

Plenty of questions like this might run through a leader's head. How does each of these questions impact on the feelings of the leader toward the project?

Does the leader like working on big projects or do they get a bit freaked out and prefer to work on smaller, less complicated projects?

Is the leader someone that prefers to be involved with activities that are perceived to be very important to the organisation? Or, do they really enjoy contributing to the smaller less-noticeable activities.

Is the leader the sort of person that gets value out of working with big budgets? Do they like the flexibility and range of choices large amounts of funds bring them?

Or, do they prefer smaller budgets? Do they feel that big budgets introduce too much chaos due to larger team sizes and more challenging goals?

Does the leader like the challenge introduced when there are hard constraints and restrictions on the money available to a project?

Does the size and scale of the project line up with the type of work the leader wants to be involved with for their own career?

Or is it completely the opposite type of project that makes them question their involvement?

Regarding the profile of the project, if it is indeed high profile, does this leader want to be in the spotlight? Do they like the idea of working with something where a significant audience will be watching?

Or, would the leader prefer to work on a project that is relatively less well-known and more behind the scenes?

Level of Risk for the Project

- Is this project high risk, average risk, or low risk?
- Would using agile project management make a positive or negative contribution given the level of risk?

In consideration of the leader's feelings in this aspect we should find out how the leader feels about risk on projects. Are they comfortable with risk?

Do they like the challenge of leading high-risk projects? Do they have experience successfully delivering projects with high risk? Or, would this project be too simple, i.e. low risk, and therefore not challenge them enough?

At the same time, given their previous feelings about project management and agile projects, how will the leader feel about this project being run using agile project management?

Relationship with Other Project Stakeholders

- What sort of relationship do the leaders have with each other and with other project stakeholders?
- Is there strong connection, trust and collaboration?

In an agile project, there will ideally be regular communication between key project stakeholders.

Recall those cultures from Chapter 5, particularly *work together daily* and *speak to each other in real-time*. If there is a good relationship amongst the leaders with their stakeholders, then you are more likely to see these necessary cultures emerging.

If the leaders *play nice together*, then there is also a chance that the agile approach will work.

Relationship with the Project Manager

This aspect of leadership is one of the most important and worth emphasising.

Put it this way: the role of project manager is to deliver the project for the leaders. The project doesn't belong to the project manager, it belongs to the funder, the owner and the leaders.

With that in mind:

- Do the leaders and the project manager get on well together?
- Is there obvious trust between each other?
- Can the leaders rely on the project manager to deliver the project for them?

Think about what might happen if the leaders don't get along well with the project manager, or they don't trust each other. What do you think might happen if the leaders feel they can't rely on the project manager?

If there is something lacking in the relationship it will often come into play when the leaders and the project manager need to work together. It could be detrimental.

Take a simple example: an issue occurs on the project that requires some form of attention and possible assistance from the leaders.

In one scenario, there is a good relationship between the leaders and the project manager, i.e. they trust each other, and the leaders feel they can always rely on the project manager to do the right thing on their behalf. In this case, the news about the issue is likely to be met with a positive problem-solving attitude and may be resolved without much fuss or concern.

In another scenario, consider a poor relationship with lack of trust. The leaders might feel that their usual unreliable project manager is coming to them with yet another problem and needs to be bailed out of trouble. You might find leaders in such a scenario are unwilling to lend their assistance as willingly or positively.

In addition to this, the perception of the relationship between the leaders and the project manager will have a flow-on effect to other project people. In particular, how the project team interacts with the project manager often reflects the leaders feeling in this context. The end-result then flows onto the outputs that are delivered and the subsequent achievement of target outcomes and benefits.

It is important that the leaders and the project manager have a good working relationship. Like any partnership, trust and reliability are critical.

Transparency of Status Reporting

Project reporting in this context means regular updates, via the project manager. The report goes out to the leaders and other interested stakeholders. It summarises the overall progress of the project and any current risks or issues people need to be aware of.

Transparency refers to how honest and objective the information included in status reports is. That is, an accurate and factual snapshot of all aspects of the project versus a subjective opinion that omits various details.

Why is it important? Once again, it comes down to trust. Additionally, you can sprinkle in a bit of attentiveness, i.e. helping the leaders be aware of current risks and issues early so they can get ready on their side to help deal with things.

Imagine a case where the project manager is not so inclined to share bad news with the leaders. Let's say this project manager prefers only to share the good news and tends to hide most other news. What do you think will happen on an agile project where this is happening?

The leaders won't be aware of the feedback coming in each time period unless it is positive. The leaders won't be aware of any risks or issues. They won't be aware of the true progress the team is making unless it lines up with the schedule estimates.

Now, what do you think will happen when, inevitably, the project manager needs the assistance of the leaders? The project manager will most likely have to visit the leaders, tell them about an issue and then ask for their help.

The leaders will obviously discover that the issue has been building for quite some time and there were several other indicators along the way. The leaders will also be caught off-guard, i.e. un-prepared.

Will they be happy? Probably-not.

Will they be interested in helping? Begrudgingly, maybe.

Will they have doubts about this project manager? Absolutely!

Such a scenario can be avoided entirely with a few simple steps:

1. find out what level of status reporting detail the leaders on a project prefer to receive
2. confirm how and when the leaders wish to receive the information
3. report the facts, honestly

The great thing about agile projects is that there is an opportunity to report status much more often and regularly. You can line up status reporting with the time-period iterations such that everyone on the project shares the successes and the issues equally.

Hardly any projects run smoothly all the time. By its very nature, a project is unique and therefore you won't be able to predict everything that will happen. Use the iterative nature of agile project management to your advantage and share the *true* project status at regular intervals.

Adherence to Schedules

A key aspect to agile projects is the use of specific intervals of time. This is the plan-output-review cycle we mentioned earlier and each cycle has its own timeframe interval.

If you have a leader that is in-tune with this approach and will work with the project team to stick to the schedules you will find that things will run way more smoothly. If, on the other hand, there is a leader that isn't so fussed with working to schedules, then that is where agile starts coming undone. Let's look at why this is the case.

Agile projects are most effective when the project team gets into a groove so-to-speak. In other words, their output rate is maximised and the team members reach a state of *flow*.

Recall in Chapter 5 how we talked about sustainable work practices? Working to a predictable schedule with well-defined timeframe intervals is one aspect of this sustainability.

Project team members in an agile project should know exactly when the next iteration will start, when it will end, and what they are attempting to produce within that timeframe. Having this level of certainty helps them estimate, plan and self-organise their work so they can do the best they can towards achieving their goals each time.

Where a leader isn't too concerned with finishing each iteration on time, you occasionally see them asking for things such as:

- increasing an iteration by a few days to get an additional output or two completed;
- changing a schedule to meet some other external deadline;

- modifying schedules to fit in with their own availability.

Such behaviour introduces uncertainty and upheaval into an agile project. For such a leader, you might help them become comfortable with schedules by pointing out that they end up with the following benefits on an agile project:

- greater certainty of output delivery since project teams improve their own ability in estimating output rates;
- regular, predictable releases of finished products;
- improved feedback from project stakeholders since they know exactly when they are expected to participate;
- more accurate estimation of project costs and less variability with actual expenditure.

Sensitivity to Modification of Plans and Priorities

Historically, there was a belief for a long time that the best way to run a project was to put in significant amounts of effort up front on things such as:

- specifying as many details as possible regarding requirements, priorities, solutions and designs;
- producing detailed estimates about effort, time and cost requirements;
- developing a *predictive plan* of how the project would run.

After all, with such great information what could possibly go wrong? How could the project not succeed? Depending on the background of your project leaders, you may find that some still find this prediction approach to project management as the one they feel most comfortable with.

You can empathise with a leader in this regard. A leader with responsibility for large budgets and delivery of outputs critical to an organisation would be quite at ease with a project team that told them exactly what they were going to produce, why, when and how. That sort of information helps justify decisions made to kick-off a project or to guide it in a certain direction.

Well… there are plenty of examples over the years, in many different industries, of projects that didn't go to plan. The evidence suggests that predictive planning doesn't work in all scenarios.

Agile project management arose in-part to deal with these sub-optimal results. The folks at the start of the agile project management movement realised that teams don't always have the right solutions up-front. They noticed that regular delivery of outputs, along with enhanced stakeholder feedback and adjustment of priorities, ended up producing better results overall.

Don't get me wrong. There are indeed many projects where it is critical to specify and plan as much detail as possible up front. In certain industries, it would be absolute madness, i.e. highly risky, to do otherwise. In these cases, a leader that was comfortable with predictive planning would have the correct strategy.

But in other situations, it might be helpful for the leader to acknowledge that it could be OK to adjust plans and priorities along the way. In these situations, when the project stakeholders start seeing the outputs, they get a better idea of what they truly need. The project team can respond accordingly.

Degree of Comfort Releasing Partial Sets of Outputs

Launch parties, promotions, fanfare, media releases, interviews, radio shows, tv spots, newspaper articles, online traffic spikes, social media spotlight, photo opportunities, etc. This is what you get when you complete a project that delivers all its outputs in one go toward the end in a *big-bang* manner.

What leader could resist the allure of all this! Apologies, I was being a bit facetious there. That's not the only reason some leaders like projects to wait until they deliver everything in a single release. Other genuine reasons include:

- not wanting the project to come across as uncertain or lacking in relation to solutions;
- wanting to avoid regularly *annoying* stakeholders and customers with new updates and new solutions;
- needing to balance operational capacity of an organisation across several projects and associated activities;

- valid situations where a set of outputs can logically only be released together, in full, at once.

A project leader that is aligned with the big-bang approach to releasing outputs may not be comfortable with the agile approach.

In agile projects, outputs are released gradually and successively as certain capabilities are completed. There is no waiting until the end. Once something is ready, it is released to the project stakeholders. The eventual full-set of capabilities for a project output might be delivered gradually over time.

A leader with feelings of discomfort around output release frequency could be helped by considering the following. Gradual, iterative, successive release of outputs provides:

- reduced risk of something going wrong with one or many of the outputs;
- less complexity of interrelated outputs due to fewer new interfaces with each other;
- less details for stakeholders to understand each release;
- more targeted feedback from stakeholders focused on higher value issues. For example, stakeholders may be less inclined to get hung up on minor issues that they know will be addressed over time and instead focus their attention on features that would make a big difference;
- increased opportunities to respond and adjust to stakeholder feedback while the project is still running.

Keep the Leaders in Mind

The topics we have considered in this chapter are important to leaders. There is no right or wrong way to feel about something when it comes to managing a project. It comes down to previous experience and the environment which a leader has come from.

It should be kept in mind that feelings tend to be exacerbated as they group together. In other words, the feelings covered in this chapter feed on each other and become accentuated.

A leader that feels positively toward agile will be a massive asset to a project.

If you find that your leaders are leaning more toward the opposite, ask yourself what you can do to help the project.

A leader's opinions are genuine and should be respected. Get to know more about them. Consider what steps, activities and behaviours you might be able to demonstrate that would greatly assist a leader in executing their responsibilities. Any concerns they have could be an opportunity to learn about areas you could strengthen for the project.

Look after your leaders. Learn from their feelings about agile projects and they will look after you.

It's not just the feelings of the leaders we need to think about though. There is also another special group that we should consider. In the next chapter, we think about the people that are responsible for the bulk of the work on the project.

Chapter 7
Take Care of the Team

There are mountains of research papers, reports, courses, coaches, studies, programs, even entire university degrees, etc. dedicated to the topic of motivating factors for project teams.

However, this book is not a scientific study, it is a *guide* to agile project management. Therefore, the material in this chapter is a set of real-world ideas, presented through a conversation.

Why go to all the trouble of making this distinction? Because when you work with people in agile projects, you need to be real. Whether you are in management, a general project stakeholder or you are working as a project team member yourself, you can't view project team members as subjects in a scientific study.

When it comes down to it, the project team is arguably the most important group on an agile project. They are the ones that produce the outputs. They are also your colleagues and some may even be your friends. Treat them with empathy. Take care of the team.

What Does an Agile Team Want?

To help answer this question I began with insights gained during my decade-plus experience leading agile teams. I also conducted several interviews with agile project team members. The consensus is that, in the context of working on agile projects, team members feel that the following elements contribute positively toward their well-being and happiness:

- good working relationships;
- direction;
- just enough structure;
- challenging work;
- opportunities for growth;
- professional working environment;
- a clear path;
- progress and achievement.

In the following sections, I investigate each of these elements to discuss what they are, why they matter, and what you can do to help.

Good Working Relationships

Agile project management recommends that teams are relatively small and tightly integrated with the roles or skills needed to produce fully-complete outputs.

In other words, you don't see a set of disparate skill-specific crews or trades coming and going throughout the project. Instead, you see a single team staying together for the duration of the project with each necessary skill or trade represented within the team.

The flow-on effect of this structure is that the project team members are going to come from different backgrounds and could be working together for quite a long time.

As such, the relationships that the members of the team have with each other is particularly important. They need to be able to get on well together and be willing to support each other as the project progresses.

It would be quite naïve to think you could change people's personalities once they join a team. Attempting to fix relationship problems after they occur in an agile project team is unlikely to work well.

However, there are a few things you could consider doing early on when the team is being formed.

Hire Slowly

First, pay close attention to *team-fit* whenever someone new is being considered for inclusion. That is, when you are evaluating a potential candidate to bring into the team, review how they are likely to fit in with everyone else from a relationship point of view.

Do they have personality traits that complement the rest of the team? Are they self-motivated? Can they balance priorities in line with what the team needs? Are they a problem solver? Do they seem willing to chip-in and help others when necessary, even in areas outside of their normal skill-set? Do they like working with others?

If possible, you might also like to consider having the existing team members themselves be part of the evaluation process when considering new members. As part of any screening or interview process you run, existing team members could meet potential candidates and provide feedback on their own feelings of *team-fit*.

Foster Relationship-Forming

Next, once the team is in place, think about ways you can foster and encourage relationships to be formed and grow within the group.

Forget about those awkward team bonding exercises you may have heard about or experienced. You know the ones I'm talking about? Those *games* where you find out *interesting* things about someone else and share it with the group. Or, heaven-forbid, where one person falls backward and the rest of the team needs to catch them.

That sort of stuff is simply uncomfortable for many people due to how contrived and fake it is. Those *gimmicks* are not how everyday people get to know each other.

Instead, just do what you would do in real-life when you wanted to get to know someone better. Go out and arrange to share an experience together.

It could be as simple as having coffee together down at your local café. Or you could arrange a nice lunch or dinner for everyone at a restaurant. Maybe find out if everyone is interested in joining in on some sort of activity together, like a game, sport, movie, concert or similar. Whatever you do, just make sure it is something everyone is positive about.

Essentially, what you are trying to do early on is make it possible for people to get to know each other personally in a setting away from the project. Have a bit of fun together. Life doesn't always need to be so serious all the time. A good way for team members to really embrace the idea of flexibility inherent in agile project management is to loosen up a bit in front of each other.

Acknowledge That New Relationships Take Time

One thing to be aware of is the fact that no matter what you do, teams don't build good working relationships instantly. Relationships take time and they will go through a few different phases on the way.

Basically, when people are getting to know each other it takes a while for most working behaviours to be seen. Everyone has their own unique personality and over time team members will begin to see various traits within each other. Initially some traits may be a bit confronting or shocking but over time people tend to settle on ways of working well together.

There will be the inevitable little issues or quirks that pop-up during the early stages of an agile project team working together. When something happens, it helps to simply acknowledge the differences and not make too much of a big deal about it. Of course, if there is some sort of major issue, it should be handled in whatever way seems most appropriate for the given context.

Direction

It goes without saying that any project team on any type of project benefits from direction and structure. In this section, we investigate aspects of direction and structure that relate specifically to teams on agile projects.

Priorities

The central question in agile projects is *what do the stakeholders want?* In the context of the project team, this translates to *what should we work on?* In other words, given the set of possible outputs that a team could work on, what is the priority order? What should they work on first, then second, third, etc.?

In the language of agile project management, the set of possible outputs that a team could work on is called the *Backlog*. What the team values most about the backlog is having it sorted into priority order based on what the project stakeholders want.

A prioritised backlog takes away any uncertainty for a team about where they should focus their time and energy. It also makes their day-to-day jobs much easier since decisions about what to work on next are already made for them, ahead of time.

Leadership

There are no explicitly defined lead roles within agile project teams. Agile recommends more flat structures within a team. One of the benefits is that team members self-organise collaboratively and therefore they tend to experiment with new ideas more often.

However, there will be times when the project team will look for guidance or consensus. This could be in relation to solution designs or choosing between competing options or for assistance with particularly challenging tasks they are working on.

One way to support the team in this context, given the lack of explicit lead roles, is to ensure there is at least one member of the team that will be recognised as having leadership qualities. This could include things such as advanced technical knowledge, empathy and care for others, previous experience with similar types of projects, enthusiasm for achievement, etc.

Among these desirable characteristics, there is one that you really want to keep a close eye out for. That is, positive support toward agile project management. A person with general leadership qualities can greatly help a project team. If that person also happens to have positive support toward agile project management, they are an amazing asset.

I strongly encourage you to look for people that fit this description and if you find them, do whatever you can to get them on your agile project team.

Their technical abilities foster confidence within the team. Their level-head helps things remain calm during challenging moments. Most importantly, their positivity toward agile project management helps everyone else on the team see the benefits of the process.

Such leadership traits act as a kind of buffer to uncertainty and scepticism toward agile that can become detrimental to progress if left to grow unchecked.

Just Enough Structure

Nobody works at their best if they are surrounded by chaos. Similarly, too many rules can stifle creativity. A good agile project should provide a team with just the right amount of structure, so they can get on with what they do best.

Later in the book, entire chapters will be devoted to some of the topics mentioned in this section. Here is a sneak preview of some of the structural elements of agile projects and why teams value them.

Which Agile Framework?

Back in Chapter 2 we discussed the various flavours of agile. One of the first things you are going to want to know, before you hire or select people to be included in the project team, is which agile framework will be used for the project. The chosen agile framework is going to influence the composition of the team, i.e. your team will ideally be made up of at least a few people that have solid experience using that framework.

Of course, when the project team comes together for the first time, they are going to want to confirm that the chosen agile framework is still part of the plan. This confirmation should be a mere formality at this stage.

Preferred Iteration Duration

Recall the plan-output-review cycle that we introduced in Chapter 1. To be able to get cracking on the job, a project team wants to know the timeframe for each iteration of this cycle.

Knowledge of how long each iteration will go for helps team members make a self-assessment of their own capacity to deliver within that timeframe.

For example, if the preferred duration is three weeks, the team can collectively put a cap on the amount of effort they have available to provide in that timeframe.

Knowing their limit helps the team then provide more realistic estimates about what could possibly be produced.

On a more practical level, knowing when an iteration starts and ends enables team members to plan their own personal and professional schedules around the iteration.

You see, within agile there is an intense focus of effort during an iteration where, ideally, other distractions are minimised.

Team members that know when they will be needed most can plan other competing work and personal activities around those dates.

Additionally, knowing when an intense focus and effort starts and ends also helps team members be a bit more comfortable with it. In general, project teams can't sustain putting in full-blown efforts over extended periods of time. If you run too hard for too long, you eventually burn out.

A well-defined duration for your agile iterations means that there is guaranteed downtime in-between each cycle.

In an agile project, there are times to focus, produce and deliver. There are also times to ease-up in between.

These bursts of energy paired with recovery breaks in-between provide a pattern of work that is sustainable for project teams.

An Overall Schedule

In a similar line of thinking to the duration of iterations we covered in the last section, here we consider the value a project team places on an overall schedule for the entire project.

A good project manager should be able to easily calculate the amount of time available for an entire project. It is normally based on an available budget and the preferred iteration duration.

It is generally a matter of dividing the available budget by the preferred iteration duration to come up with a rough number of possible iterations available.

There is a bit more to it than that but at its essence that is the starting point.

Beginning with the rough number of iterations available, a project manager can then include other relevant blocks such as:

- project initiation and ramp-up activities at the start;
- a set of iterations making up the bulk of the project schedule in the middle;
- stabilisation, commissioning and release of outputs after the main set of iterations is complete;
- acceptance, conclusion, tidy-up and shutdown activities at the end.

This is the basis of a simple project schedule. Early on in an agile project it doesn't need to be any more complicated than this.

Having such information available helps project teams to know what activities are happening when. They can then use this information to help plan their own day-to-day work.

Team members can also use the information from an overall project schedule to align activities in their personal and work schedule.

Major Events and Activities Booked in Advance

Let's continue the theme of project teams enjoying knowledge of what is happening, when, and for how long. Next up we cover the value of having all the major events and activities for the project booked in advance.

Given our focus in this book is on Scrum, we're talking about the following major events and activities:

- Start and end dates for iterations, a.k.a. sprints. The overall schedule created earlier gave us a possible number of iterations, and now the team will want to know the start and end dates for each iteration.

- Dates of planning sessions. Prior to each iteration, an agile project runs at least one planning session. These planning sessions help the team determine what work is coming up in the iteration and how much work they will attempt to complete based on their own estimates of work complexity.
- Dates of demonstration sessions. Another agile event is the demonstration. These take place at the end of an iteration and are a way for the team to showcase their progress and completed outputs to stakeholders.
- Daily team meetings, a.k.a. Stand-Ups. This mechanism is a way for agile project teams to have regular conversations together about the work they are focussing on. Ideally, they usually run for a short period at the same time each day.
- Dates for retrospectives. These are the self-review sessions that agile project teams run at the end of an iteration. They help teams identify which of their work practices are functioning well and if there are opportunities for improvement.
- Other key events. Depending on the type of project, organisation, industry, etc., there will often be other major events that will take place. While these are not necessarily agile-specific activities, they are important to the type of project. Examples include quality assurance evaluations, final inspection and commissioning of outputs, handover from the project to ongoing operations management, etc.

Once this set of events and activities are known and booked in advance, the project team really doesn't have to worry about much else other than getting on with producing outputs.

Challenging Work

One of the reasons team members enjoy working on projects is because no two projects are the same.

The Project Management Institute[xiii], an internationally recognised organisation associated with the project management profession, defines a project as:

> *A temporary endeavor undertaken to create a unique product, service or result*

The key words here are temporary and unique. Project team members enjoy the challenge of producing something that hasn't been done before. Project teams get satisfaction out of producing new and unique outputs that challenge them professionally and intellectually.

They also get value out of the knowledge that each project will have a definite start and end date. In other words, project teams enjoy knowing they won't be doing the same thing day-in, day-out.

With this in mind, there will be times in an agile project where you may need to consider rotating responsibility for certain tasks, so they don't end up becoming monotonous.

Examples include hosting or chairing planning sessions, being the scribe (i.e. the person that takes notes), running the daily stand-up sessions, etc.

A good way to ensure that nobody in a team ends up getting lumped with repetitive and non-challenging work is to introduce a voluntary rotation system. That way teams can decide together how to share the load in a free and equal manner.

Opportunities for Growth

Following from the previous section, project team members that enjoy a challenge would also appreciate opportunities to improve themselves professionally. In other words, project team members value being able to advance in areas including:

- work experience;
- professional skills;
- exposure to new technologies;
- challenging situations;
- problem solving;

- organisational relationships;
- social interaction.

Agile projects enable growth in these areas. However, there can be a tendency for certain people to get relatively more opportunities than others.

Depending on the types of personalities present within a team you might find that some team members dominate over others, whether they explicitly mean to or not.

The nature of agile projects is that teams support each other and decide amongst themselves who gets to work on what each iteration.

If you have someone with relatively strong technical skills, the team might naturally defer to them when it comes to working on the more difficult tasks.

Similarly, if you have someone with a relatively dominant personality they could end up taking tasks that suit them over others.

Like we mentioned in the previous section, it would benefit the team to acknowledge this situation. One solution is to explicitly build in opportunities for sharing and rotation around tasks which are valued growth opportunities.

Professional Working Environment

It goes without saying that supporting a project team with modern tools and efficient workspaces enables productivity. There are a few agile-specific aspects of this to consider.

Co-location

Recall back to Chapter 5 where we spoke about the cultures important to agile projects. One of those cultures encouraged people to work together daily and another referred to people speaking to each other face-to-face.

Communication isn't just something that is taken for granted. Communication is a primary driver behind the results delivered by an agile team. The idea is to have the most effective and efficient form of communication possible.

Real-time conversations that happen in-person enable a high level of productivity. If a team member needs an answer to a tricky question, wouldn't it be better if they could turn around and ask someone nearby to them?

To answer this, think about what normally happens when people are physically separated. Either they need to make a phone call or send some form of electronic communication, such as email or instant message.

Depending on what the person on the other end is doing, there may be a delay in the reply. There could be a clarification required, which could result in back-and-forth and so-on. In some scenarios, this non-face-to-face communication could take a few minutes but when it is particularly bad it could result in delays of several days.

In a co-located agile team, communication delays are reduced to mere seconds.

An additional benefit of being co-located is that teams can pick up on non-verbal cues from each other. Seeing body language helps the project team to empathise with how each other is feeling at a given time. This helps form and foster relationships. A team that is in tune with each other is one that works better together.

"But hang on a minute", you might say to yourself. "Sometimes people need to focus. People can't be in each other's' face all the time". This is a correct statement. Two of the side-effects of people working closely together in groups are noise and disruption. We deal with this on agile projects using work zones.

Work Zones

Agile project teams have a variety of interaction levels with each other and with their stakeholders. As such, an ideal working environment is one that provides various locations that support the necessary interaction levels.

There are times where the entire project team needs to speak to each other and collaborate on the production of outputs. This is generally during their daily stand-up meetings, but it can also include joint solution design sessions.

Important aspects to consider in this context are that the team can comfortably fit in the same vicinity, that they have access to their working materials and that they can speak freely.

If you are in a small organisation where the project team members are the only people in the area, then that is not a problem. If, however, your team is part of a larger organisation and there are other people working nearby then you might consider finding a separate location where the team can go and be noisy together without disrupting others.

There are other times when individual team members benefit from quiet uninterrupted focus time. Depending on the resources of your organisation you could support this through a few individual offices or noise-shielded desks.

In agile projects, you will also see regular demonstrations of project deliverables to stakeholders. A good way to support this is with a big meeting space that is suitable for larger groups of people to physically show and review actual working outputs.

The common element in the above two sections has been team members working together in the same location. Sometimes this just isn't physically possible. What do you do then?

Modern Communication Technology

Alright, reality time. I have been rabbiting on in the last two sections about co-located teams and work zones for team members to interact with each other. In today's working culture, there is an additional requirement.

There has been a strong push from around 2010 onwards to support remote knowledge workers. That is, team members who provide predominantly desk-based work contributing from whichever physical location suits them best. They could be at their home, commuting on their way to the main building, at an office in another city, at an airport waiting for flights, etc.

I have managed agile projects for over a decade where team members were scattered across many locations around the world. These teams did just as well working remotely as they would have done if they were in the same physical location. The trick that made this possible was a modern set of communication technologies.

If you are going to bring together an agile project team in a virtual location, supporting them with communication technologies is the way to do it. However, there are a few considerations to keep in mind. If you're not careful, you could end up with a group of separate individuals that communicate when it suits them rather than the collaborative and supportive team that agile aims for.

First, keep in mind that there is a general phenomenon whereby communication technologies stick. That is, the first few technologies that everyone uses in projects tend to be the ones they keep using. Research and experience has demonstrated that once a team is comfortable using a set of communication technologies, it is very hard to get them to switch to something else.

I would strongly encourage you to focus on communication technologies that mimic the way project teams normally interact with each other in real life. That is, real-time face-to-face conversations. Refer to the discussion in the previous section on co-location for the reasons behind this. Such technologies include:

- Telephones. Standalone is OK but these days you can integrate phones with computers to get a richer interactive experience that makes use of the next item.
- Web cams. For a few hundred dollars, you can get ultra-high-definition video cameras that easily plug-in to most computers. Some machines and devices are lucky enough to have these cameras built-in.
- Web conferencing. The software in this space has matured nicely and today you can easily bring together large numbers of people in a virtual room. In well-featured products, you can see a live video of each participant, share your desktop screen, sketch diagrams on a virtual whiteboard, etc.

There will be times on an agile project when you bring groups of people together. For example, to host demonstrations or planning sessions. If your organisation can afford it, it would be a sound investment to kit out any shared meeting rooms with the same sort of communication technologies. Additionally, meeting rooms need to have large screens that everyone can see no-matter where they are seated. The rooms should also have a set of high-quality microphones that can pick up conversations.

Such equipment can be quite expensive though so if the budgets are thin, the fall-back is for everyone to connect in from their desk.

I would suggest that when it comes to deciding which to use, i.e. people in rooms or at their desk, pick one or the other but not both.

Having a large group of people in a room while others dial-in from their desk doesn't work too well. The people in the room naturally tend to have side-conversations and this generates a lot of noise that the microphones pick up. Trying to focus on a single conversation when you are listening to it through a speaker or headphones is quite hard. All sound is combined so you end up hearing several other voices and background noise together.

If you really have no choice but to combine people connecting from rooms and desktops, then be sure to politely explain the issue to the people in the room before each session and ask them to limit conversations to one person at a time.

You might be wondering at this point why I haven't mentioned other pervasive communication technologies like email or instant messaging or group/team chat. Again, refer back to the discussion on co-location. These other communication technologies are known as text-based asynchronous messaging tools. In other words, one person types a message, it gets sent, they wait, someone else types a response, sends, they wait, etc.

In agile projects, you really do try and foster people *speaking* to each other whenever possible. Active communication that involves voice and body-language interaction is preferred over passive communication.

There is still a place for written communication on agile projects. In the next section, I discuss some of these options and other ways agile project teams can be supported with the tools they need for the job.

The Right Tools for the Job

Your project is going to require all sorts of tools so people can do the work to create outputs. Whatever you do, make sure the team gets access to tools that enhance their abilities and efficiency.

You will need to make cost-benefit trade-offs. Of course, most organisations have limited budgets to spend on such tools. But in doing so, consider the impact these decisions will have on the productivity of the team.

Additionally, consider the team member's preferences when you make tool selections. A tool that is wonderfully efficient and makes people more productive isn't much use if the team members hate it.

A Clear Path Forward

When an agile team goes on the journey of completing a project, they will occasionally encounter impediments along the way that slow them down or even prevent them from moving ahead. These types of issues are known as *blockers*.

Understandably, to be at their most efficient and best, project teams want blockers removed. Such blockers might include:

- waiting for a task to be completed by someone else before being able to complete another dependent job;
- requiring approval to undertake an activity;
- unclear requirements and features for a project output about to be worked on;
- lack of knowledge and experience regarding where to start on a particularly challenging problem;
- broken or inefficient tools and materials.

This is where all that effort we put in co-locating the team and ensuring they communicate to each other in-person starts providing a return to the group.

Ideally, an agile team should be encouraged to speak up whenever they hit a blocker. That way there is a chance that someone in the group might be able to assist in resolving it.

What would make everyone on a project comfortable to speak up whenever they encounter a blocker?

A culture is required where requests for assistance are viewed positively and everyone in the team steps up to help whenever it is needed.

This culture should also be fostered in the opposite direction. That is, project teams should be encouraged to look out for signs of someone struggling with an activity. If it looks like someone needs help, project teams should reach out and offer it.

There will be times when certain blockers are outside the ability of project teams to resolve independently. In a well-functioning agile project, you see other project people also getting involved to help when they need to.

What you are striving for on an agile project is a community approach toward making progress. In other words, removal of those artificial barriers of *us vs them* that often emerge in organisations. Aim for everyone involved in the project to look out for one another.

It would be fair to say that some of the suggestions made in this section are easier said than done. Human behaviour is often a conundrum. However, there is one thing I'm quite certain brings people together. In the final section of this chapter we look at the importance of success to the wellbeing of agile project teams.

Progress and Achievement

There is plenty of research out there today confirming that financial incentives are not the primary motivators for people in jobs. I know, crazy right? Isn't making money one of the main reasons people start jobs in the first place?

Forgive me for being facetious. I am a stern believer in progress and achievement over financial rewards. Why else would I have stepped aside from the lure of fee-based project services to write books like this?

Believe me, there is far less financial reward in writing and self-publishing than there is in consulting. My main driver is creating something and sharing it so that others can have a chance at the success I have achieved with agile projects.

Do you know what really motivates project teams? Meaningful achievement with their peers. Project teams love completing jobs successfully together. It builds a sense of comradery and purpose.

Team success is also one of those wonderful cases where the combined result is greater than the sum of its parts. That is, the positive outcomes the project team gets as a whole is greater than if there were an equivalent set of wins for each individual alone.

Much of this comes down to evolution and human nature. We are hard-wired as homo-sapiens to be part of groups and to achieve success together. Back in hunter-gatherer times, only by sticking with the group and successfully finding food were we able to survive and pass on our genes to the next generation.

Putting this in the context of agile projects, it is important to structure project team activities so that success is not only possible, but likely.

The primary feature in support of progress and achievement on of agile projects is the iteration.

At the start of an iteration, a project team estimates and plans which outputs they will target.

Within an iteration, agile project management encourages teams to achieve each day. They check-in during the daily stand-up to review their progress for the past day.

At the end of the iteration, the project team demonstrates what they have completed to the interested stakeholders for the project.

Over a series of iterations, the team moves closer and closer toward the end goal of completing the project.

As you can see progress and achievement is built-in during all stages at both a micro and a macro level.

It is important that activities around the iteration are undertaken in ways that foster progress and achievement.

Thus, one thing you want to keep in-tact in an agile project is the iteration framework. Make sure iterations are treated with respect.

However, having just pointed out how important proper iterations are to progress and achievement on agile projects, there is something to be aware of.

There is a quirk of project teams working in iterations that often impacts progress and achievement. This oddity is the tendency for people to under-estimate the complexity of producing outputs and over-estimate their ability to complete work in each timeframe. Without a single exception, I have seen this show up on every single project I have managed, regardless of the industry, profession, experience or skills of the people involved.

If you are not careful, project teams could find themselves setting stretch targets for themselves and then regularly failing to achieve such targets each iteration. If this starts becoming a pattern, then you begin to see psychological impacts on individual team members. A feeling of failure spreads fast and once it sets in there is a cascading effect on various other aspects. Motivation levels drop, productivity falls, outputs need to get de-scoped. It becomes a spiralling and self-defeating cycle.

In a later chapter, I am going to spend considerable time discussing this issue of progress and achievement in much more detail. It is so very important that it deserves more than just a cursory glance.

Smells Like Team Spirit

In this chapter, we have considered the importance of caring for the feelings of agile project teams. The underlying premise of team spirit is what this chapter is all about. Now that you understand ways in which agile projects can foster or inhibit team spirit, you can do your best toward contributing positively and enabling others.

In the next chapter, we continue the theme of team and talk about the recommended agile approach of self-organisation.

Chapter 8
Self-Organisation

Recall back in Chapter 5 where we introduced an Agile Manifesto item that stated the following:

The best architectures, requirements, and designs [i.e. solutions] emerge from self-organising teams.

In relation to this we talked about micro-managing and the idea that project teams generally produce their best work if they are trusted and left to get on with the job. At this point, I invite you to pause for a moment and think about the implications of these ideas.

In this chapter, we discuss the concept of self-organisation in the context of teams. We go through what it is, how it works with agile project management, how to implement it effectively and things to watch out for.

Before we get into this, however, I'd like to let you in on a fairly obvious principle of the project management profession. Keep this in mind as we progress through the chapter.

The best type of management activity is one where people don't feel like they are being managed.

What is Self-Organisation?

A good way to introduce the idea of self-organisation is through a basic example that most people would be familiar with. Think about a group of adults and children. That group could be anything you want, say a school, or a sporting team. It doesn't matter, just assume that there are a bunch of kids and some adults in charge of them.

In most settings like this, you will see the responsible adults managing kids in a hierarchical manner. For example, teachers usually set the lesson plans to fit with the curriculum. In another example, coaches usually set the boundaries for their kids around activity times and rules of the game, etc.

Let's assume this group we are imagining has quite a few children, say more than five, that are relatively close in age. Next, let's say the kids have been given a task to complete by the adults. Make it a fun one.

Assume that it is the end of the week, the adults are tired, and they need an hour or so of uninterrupted time to write up a newsletter and report about the kids progress for the week. The task set for the kids is to simply go outside together and play for the next hour, so the adults can have some quiet time to focus.

Initially, the adults may direct in some way. For example, they set boundaries about where the kids can or cannot go, how long they can play for and what equipment they are allowed to use. Mainly though, there will be a long stretch of time where the children are left alone without the adults.

During these moments, kids will do what kids love to do: play games and get up to mischief. An interesting thing happens when groups of children get together to play. First, their imagination runs wild and they start deciding among each other what to play, how to play it, what the rules are, where it will go next, etc.

Next, there is the inevitable establishment of a pecking order. Some kids love trying to tell the others what to do, other kids go along with it, while others come up with crazy ideas that gets everyone else excited.

Together, kids in such situations usually end up settling in to some sort of equilibrium regarding the social structures of the group.

Additionally, and most importantly, they often end up having a great time together. They are free to come up with all sorts of wild and wacky games and crazy antics.

This is an example of self-organisation at its most finest (and most enjoyable). No pesky adults getting in the way telling the kids what to do. The kids get the freedom to invent whatever on-earth they want to.

Sure, the adults may check in on them occasionally or help them resolve a problem like recovering a ball that got stuck up a tree but, overall, they usually have a wonderful old time.

At this point, let's pause the example for a second to provide a definition.

Self-organisation is the spontaneous arranging of components within a system, in a way that is purposeful, without the control of any components outside the system.

Back to the example of the kids. The group of children together is the system. Each individual child is a component within the system.

Anytime they get together for a play session, they naturally organise themselves and randomly decide what they want to do. That is, they spontaneously arrange.

They are certainly purposeful. Have you ever had a conversation with a child that is telling you about the amazing game they just finished playing? Oh, there is purpose in there without question.

Lastly, the group of kids is without the control of any components outside the system. That is, there are no adults, and they *love* it!

Ok, are you still with me? Let's take this understanding and apply it to a different example.

Imagine a new scenario where an organisation is running a business project.

In this example, it is a modern company with project team members, stakeholders, appointed leaders and managers. The goal is to produce outputs that meet the needs of the organisation, the project and the project stakeholders.

Here is where our definition comes into play. In this new example, the project is the system and the project team members are components within the system. The leaders and managers are components outside the system.

In some settings, the managers and leaders of the company will be tightly coupled with the project team and they will decide who does what, when, how etc. This is the old traditional command and control style.

Let's say, however, we have a situation where the managers and leaders only get involved with the team as far as setting the direction for the project. That is, they set the goals of the project, the timeframe and the available budget but nothing else.

Assume also that once this direction has been set, the organisation leaves the team to get on with the job. No leaders or managers are embedded within the team. Nobody tells the project team members how to do their job. They are simply given the boundaries and constraints and then allowed to get on with it.

There are no pre-defined hierarchies with the project team. Everybody in the team is simply part of the group and they all get a say in how they are going to deliver the outputs required. They work within the given boundaries and constraints.

In our example, the leaders and managers occasionally step in to ensure the team is progressing. They may also assist with resolving issues that come up from time to time.

This scenario we have just described is self-organisation in the context of a business project.

What Self-Organisation is Not

The first thing to note is that self-organisation does not mean self-directed.

Did you notice that in both examples, there was initially a set of external guidance and direction provided?

That is, the goals of the group were given along with the boundaries and constraints.

In the example of the children, the adults guided their activities. The kids were told they were to go outside and play. They were given boundaries, e.g. "stay in the backyard, don't go out the front or beyond". They were also given a set of constraints, e.g. "you can use the balls and equipment but please stay out of the shed and don't play with the tools and machines".

Similar guidance and direction was provided in the business project example. "Here are the goals of the organisation and project, we need a set of outputs that deliver on this and meet the needs of the stakeholders". The team was also provided with constraints, e.g. "there is a fixed budget available and we need the work completed by the end of the year".

The second point to note is that self-organisation does not mean self-managed.

In the example of the children, the adults would occasionally check-in to see how everyone was getting along and to know what the kids were up to. The adults would also step in when necessary to help solve problems like recovering equipment or breaking up arguments.

Similarly, in the business project example, the leaders and managers check-in regularly to ensure the team was progressing and delivering outputs. They would also help resolve any issues that arose.

How does Self-Organisation Work with Agile?

To describe self-organisation within the context of agile projects, let's view what is provided externally, and what the project team members are left to take care of themselves.

In an agile project, the following aspects are pre-determined:

- Goals of the project are set.
- Target outcomes are defined.
- Stakeholders of the project are known.
- Customers of the project's outputs are identified.
- Guidance is provided about what types of outputs are going to be produced.

- Available budgets and timeframes are given.
- The constraints of the job and working environment are fixed. For example, the projects outputs are built within certain logical and physical boundaries using certain available equipment.

Additionally, agile-specific structures are put in place. These include items such as the following:

- The project team follows the plan-output-review cycle.
- The project team holds a master list of possible outputs to be worked on and they keep that list in order according to the priority of the project's stakeholders.
- Project team members demonstrate their completed work at regular intervals so that stakeholders can review them.
- Governance mechanisms and reporting lines are established in association with the project team.

Once the direction is set and the boundaries and constraints are defined, the project team commences their journey of self-organisation. In particular, the team members:

- decide on what they think would be the most appropriate iteration timeframe for the project;
- pick a regular time for their daily stand-up meeting;
- regularly interact with key project stakeholders to determine and adjust priorities of the outputs that will be worked on;
- decide on solution and output designs together;
- estimate on what outputs they might be able to complete in each iteration;
- self-assign and adjust who works on what according to the most relevant skills associated with particular outputs;
- manage their own workloads;
- ask questions and attempt to find relevant information themselves, i.e. they don't wait for it to be fed to them;

- attempt to respond to issues and solve problems together within the team first;
- work together at the end of an iteration to integrate and demonstrate their completed outputs;
- reflect on their own performance as a self-organising team and adjust for improvements as necessary.

If you head back to the childhood example earlier, you can see why self-organisation is so effective.

By our very nature, when humans work in groups we are wired to want to work together independently. In other words, we like working inside the group but we don't necessarily like being told what to do by people outside the group.

With self-organising agile teams, coaching, guidance, assistance and direction is fine. But specific instructions and micro-managing? No thanks, stay out of it.

With self-organisation in agile projects, the team members are trusted as professionals to come up with the best solutions they can given their environment and the constraints they face.

The work in such a setting becomes very satisfying for the individuals within the group. The group is able to set its targets each iteration and then work toward them.

Of course, they don't do this in isolation. Ideas for solution and outputs designs are verified with key project stakeholders before work starts.

At the end of each iteration, the team gets to deliver the outputs they have completed. Key stakeholders review the work, provide feedback, talk about what they like and give suggestions for where they would like the work to go next. It is an extremely collaborative process that triggers positive responses for everyone involved.

If you were to sum up the concept of self-organisation for agile teams into a few words, they would be: positive guidance, recommendation, coaching, support, trust, freedom, collaboration and achievement.

Things to Be Careful of with Self-Organisation

Time & Patience

Growing a self-organising team takes time. The fact that self-organising teams have a flat hierarchy structure, i.e. everyone in the team is equal, means that they need to reach their own natural state of equilibrium. It literally is a case of the group figuring out among themselves how they will be productive together.

The beautiful thing is that once everyone in a self-organising team eventually does settle into a comfortable pattern, you normally see reliable, high-quality, regular delivery of outputs.

The solutions they produce are highly innovative reflecting the freedom they feel. When a self-organising team gets into a state of flow together, it is a wonderful site to behold. Just don't expect it to happen immediately.

Culture Shock

If the group includes team members that are conditioned to active management, or worse, micro-management, there may be a period of initial confusion about what self-organisation means. If someone has been told what to do their entire career, or if they have always needed to check with superiors, this new world can be confronting.

Team members that are used to having others organise their work often have a common initial response. They feel that the leaders and managers are somewhat disconnected.

There can also be a feeling of being lost at sea due to the need to actively come up with solution options themselves.

This is only natural for someone taken out of their normal environment and placed in a whole new world. Once most people in this situation come around to understanding what self-organising means, the realisation is liberating to them. The opportunity to experiment with new ideas is like watching a caged bird fly free.

Self-Organisation May Not Work for Everyone

There is a general scenario that you want to watch out for with the effectiveness of some members of self-organising teams. At a broad level, it is simply a matter of capabilities not fitting the environment.

For various reasons, some people just don't perform best in self-organising teams. For example, some people may genuinely enjoy more predictable styles of work without the continuous challenge of problem analysis and flexible priorities that comes with producing outputs on agile projects. Another example is people that deliver their best work alone without the regular interruption and interaction of others.

The point to take-away from this is that everyone is good at something in their own way. There is no *wrong way* when it comes to people's abilities and skills. Everyone has their strengths. It is just that some people's strengths simply don't match with the environment and requirements of self-organising teams.

How Progress is Measured

In the old traditional style of project management, folks in charge of the project compiled predictions and ideas together at the start and would measure progress against those plans. For example:

- Outputs and associated work packages were specified.
- Required durations were estimated for each work package.
- Complexity and associated number of people required to be assigned to work packages was assumed.
- Schedules were created and milestone dates were set.

With self-organising teams on agile projects, this old approach doesn't apply. Instead, the constraints set in advance are time and budget. It is up to the team to produce what they can within those constraints.

Those of you familiar with project management concepts may recognise this as an implementation of the old iron triangle. Remember the one? On projects, you have three constraints of time, cost and scope. You can possibly fix up to two of them, but this comes at a cost. The remaining one must be flexible. You cannot fix all three.

In self-organising teams on agile projects, you are structuring the project such that there is fixed time and budget available. Therefore, scope must be flexible. Self-organising teams have the freedom to produce outputs within those constraints.

Once the team members have been selected and the team is formed, it is important to note that their capabilities are thus set. They will deliver whatever they are capable of and have capacity for.

Members of a self-organising team will produce as much as they can in the time available and come up with ideas to the best of their abilities. Nothing more, nothing less.

This style of work can challenge some people, especially if they have an unconscious bias toward the old traditional ways of predictive planning.

People, especially managers, that think in terms of schedules often base their ideas of progress on whether the team gets through the planned work in the set timeframe. Delivering less outputs than the expected plan is sometimes seen as an issue, or failure, by people with this view.

The conclusion is, don't judge the progress of a self-organising team in agile projects on the quantity of work they deliver.

The primary goal of self-organising teams on agile projects is to ensure that the outputs produced meet the needs of the stakeholders, not the needs of a calendar.

Judging self-organising teams on how much work they get through in each timeframe is the quickest way to destroy morale. It is the wrong measure and focusing on it only serves to encourage project stakeholders to look for issues.

Instead, support self-organising teams by trusting them to deliver. A team of professionals will produce what they are able to. Their outputs will reflect their skills, capabilities and capacity. Don't make the mistake of judging people on time if you have explicitly guided them to focus on the quality of fitness for purpose.

Over-Engineering Solutions

Based on the last issue, you might think to yourself "if a project team is focused on the quality of solutions, and not the schedule, isn't there a risk they could spend time over-engineering solutions?". You'd be right.

In agile projects, there is a guiding principle that project teams focus on gradually producing a series of components that combine into the overall set for whatever is being built. The collaborative nature of agile is such that the key stakeholders of the outputs being created are encouraged to engage closely with the project team. This opens up a challenge.

It is very tempting for somebody having something created for them to ask for more. A little change here, a new design element there. "It's only a little modification right, we can do that can't we".

At the same time, in self-organising teams, there is often the desire to deliver well. There is a tendency of self-organising teams to naturally respond positively to key stakeholders. After all, who wouldn't want to do the best for a key stakeholder that is positively reinforcing and validating outputs a team is producing.

The risk, and one that does indeed occur from time-to-time, is that self-organising teams end up over-engineering their solutions. That is, they create more than they need. Their solutions can sometimes include attributes and functionality in outputs that are not absolutely necessary.

In terms of the language of requirements prioritisation, they work on things that are *nice to have.*

You might think "what's so bad about that?". After all, isn't it important that an agile project meet the needs of the stakeholders? While that is true, if the project team focusses too much on each output they produce, they end up producing less outputs overall.

It is a matter of simple arithmetic. If you have a fixed amount of time, say a weeks, to create outputs and the self-organising team is producing, on average, b outputs per week, then by the end of the time they will have produced around $a \times b = c$ outputs. The lower the value for b, i.e. the longer it takes to produce each output, then the lower is c, the number of outputs in total.

One of the primary measures of success in a project is whether its target outcomes are achieved in the long run and therefore, did the project deliver its intended benefits.

A self-organising team might meet various other measures such as time, cost and quality of outputs. However, there is a bit of a problem if there are no actual benefits available from the project.

A successful project delivers benefits, not just outputs.

Let's look at this in terms of a few examples. A project that delivers half a house is pretty useless, even if the half that gets completed is top-grade quality with amazing precision. Similarly, there is no point running a project to develop a new product to take to market if the product is only half-done by the end of the project.

Guidance is OK

At the start of this chapter we mentioned that self-organising teams were not self-managing or self-directing. This doesn't mean there is no place for management or direction.

They say it takes a village to raise a child. Well, in the same way, it is OK for leaders and managers to get more involved in agile projects from time-to-time.

If there are signs that self-organising teams are experiencing some of the issues we have identified here, then it is entirely appropriate for leaders and managers to get involved more actively.

Good leaders and managers are adept at identifying when it is appropriate to provide guidance and coaching to a self-organising team. If there are issues occurring, there is nothing wrong with stepping in for a while to provide suggestions on getting things back on track.

Wise leaders and managers will consider how they get involved. They will be aware of the strategies behind the team remaining self-organised. They will take care to not come across judgemental or discouraging. They will respect the autonomy of the self-organising team.

Remember, the goal behind self-organising teams in agile projects is to enable the freedom that results in more innovative solutions.

Up to this point, we have considered the leaders and thought about the team. It is now time to cast our gaze outward and contemplate the rest of the people involved in the project.

Chapter 9
Support the Stakeholders

Have you ever worked closely with someone that is just amazing with people?

You know the type. They're the ones that if they were hosting an event or a party they would go to great lengths to think carefully about the individual needs of everyone on the invite list.

They would make sure the music selection didn't play any specific songs that some people despised. They would include a few special items on the menu that they knew were all time favourites. They would ensure that *those two* were not sitting too close to each other. They would go out of their way to include some of the more shy or reserved people in ways that made them feel comfortable.

This is stakeholder management at its finest. If I am a recipient of such special attention, it just makes me feel so positive about the whole event or activity. It makes me feel considered.

Supporting stakeholders is a lot like this. It is thinking about the unique needs of everyone regarding their involvement with the agile-specific aspects of the project. It is then doing what is needed to try and meet those needs. It is about going that extra mile to keep people connected with the project and comfortable so that their experience is positive.

The goals are delivering outputs within project constraints, achieving target outcomes and generating desired benefits. Supporting the stakeholders is one of the means of achieving these goals.

Empathy

The name of the game when it comes to working with stakeholders is showing empathy. First, what is empathy? The Oxford Dictionary defines it as such…

Empathy: The ability to understand and share the feelings of another.

The word empathy comes from the Greek word *empatheia*. This is made up of two parts: em *in* and pathos *feeling*. In other words, you are literally placing yourself inside another's feelings.

Years ago, I had the wonderful opportunity to lead several projects with someone who was truly amazing in the empathy department. This person was a senior executive in a large organisation and had many years of experience under their belt.

I was in awe of this person's ability to genuinely consider the needs of others on the project. They would always go the extra length to care for anyone on the project, no matter who it was.

For example, this person would often drive a six-hour round trip just to spend the day working with members of the project team located interstate. This person knew that having a close relationship with these team members would be good for the project overall, so it was done without a second thought.

In reverse, casual social gatherings would be arranged anytime key stakeholders would travel to visit their main office. The goal was to let these travelling stakeholders know they were being considered.

A nice meal was always arranged in a relaxed setting as a way of providing some comfort to those travelling away from their homes. It was a sincere touch.

For me personally, I can recall that after the first few times I travelled away for work, the shine and excitement wore off and I did start missing the comforts of home.

I learned a great deal from this person on those projects. While the projects did have a few ups and downs, the stakeholders always went away with a positive experience overall.

Every one of those projects was successful. I put a large part of it down to the way my colleague showed genuine empathy to each and every one of the project stakeholders we worked with.

Who Are the Stakeholders?

Recall back to Chapter 4 when we introduced project people. Well, to cut a long story short, all of these groups are technically stakeholders. They each have something to do with the project. Stakeholders will be related to the project in one or more of the following ways. They will:

- work on or collaborate with the project;
- contribute to the project in some way;
- benefit from it;
- utilise one or many of the project outputs;
- be impacted by the project.

When you think about it though, we don't naturally refer to the project team as stakeholders. Nor do we consider the leaders of the project, i.e. the funder, owner and project manager, as stakeholders. These folks have their own well-defined and special roles within the centre of project activities. You could simply refer to them as the core project group.

When we use the term stakeholders we are more commonly referring to the people associated with the project who are outside that core group. That is, customers, beneficiaries, suppliers and impactees.

Why Do Agile Projects Need Stakeholders?

You have most-likely picked up by now that agile projects are designed from the ground up with stakeholders in mind. The primary directive is to create outputs that match the needs of the customers, within a given set of constraints. The goal is that customers utilise the outputs to generate the target outcomes, leading to the achievement of benefits.

The project team is the main group creating the outputs, occasionally with the assistance of other external suppliers.

All the while, the project needs to ensure it listens closely to any impactees that may exist so that the burden on them is acknowledged, respected and handled appropriately.

It would be fair to say that stakeholders are the centre of the universe for agile projects.

Agile projects need stakeholders so that everyone involved knows what to focus on producing, and what to be careful about.

Agile-Specific Needs of Stakeholders

There are many things project stakeholders need in general. For example, they need to be given ample opportunity to be consulted and to have their thoughts about the project heard.

If they are working on the project, they need to be supported in relation to the specific activities they are working on.

If they are providing some sort of governance service to the project, they need to be kept informed regarding progress and any elevated risks or issues.

The list goes on. I could fill an entire book on the topic of project stakeholder management. Instead, let's look at the agile-specific needs of stakeholders.

As an exercise, I thoroughly encourage you to become familiar with the more general elements of stakeholder management.

Be a Guide

Stakeholders don't work on agile projects full-time. They are not what you would call *agile practitioners*. As such, they are less likely to know the ins-and-outs of whichever flavour of agile you have chosen to work with.

Therefore, if you want to maximise the value available from stakeholders working with your project, it would be wise to gently guide them in the right direction along the way.

You know the saying, right… never assume anything, it will only make an ass of you and me. Don't assume your stakeholders will know what agile specific things you want them to do.

Help your stakeholders out by explaining agile specific things to them the first time you run them together. Discuss what you are about to do together and why it is important. Most of the time, something as simple as this helps give your stakeholders the context and understanding they need to work with you.

Speak Their Language

My lovely sister-in-law, Melinda, works in an area where she is often a stakeholder on agile projects. Melinda gladly offers up her own views on agile to me. Melinda's biggest gripe, is that the core group of people in many of the agile projects she has been involved with made things more complicated than necessary. The issue was the terminology that they used when speaking about their projects.

I have observed the same myself on many occasion. The use of jargon can be a bit off-putting for some people, particularly for those who are on the fringes of supporting your project.

Think about it like this. Say you have spent your whole life understanding a certain word to have one meaning. One day, someone comes along and uses that word in a way that is completely at odds with the meaning you believe it to have.

You are probably going to be somewhat taken aback while your brain tries to make sense of what was just said. At first you might simply nod along and pretend to understand what they are saying.

Once you think you can pinpoint the new meaning of that word, you will inevitably compare it to the definition you originally had. Which will lead you to the thought… "that's not what that word means!".

I can recall the first time I heard someone use the term *story* in the context of an agile project. I was a newly minted project manager and the members of my team were bandying about the word story in their conversations about upcoming work.

Initially, I had no idea what on earth they were talking about. I had to ask them to pause one day. "When you say story, what do you mean?" I asked. "Oh, stories are the outputs we are creating". Still stumped, my reply was something like "Okay, I didn't realise we were creating a fiction novel".

Eventually, I came to learn the new meaning of the word story when used in the context of agile projects. But honestly, it didn't make sense for a while. My team members were abbreviating the full phrase *user-story* and using it casually in their everyday conversations. This just confused the heck out of me, until I eventually got it.

For those of you reading that might still be struggling yourself, it is only fair of me to provide a brief definition.

A user story is a method of describing an output in terms of the way the end user will interact with it to achieve a certain goal. It goes like this: as a [particular user/role], I want [the output], so I can [achieve a goal]. For example, as a bank customer, I want to pay for things at the store using my phone, so I don't have to carry around cash or cards in my wallet.

Anyway, back to the point. Some words and phrases in agile projects are used in ways completely different to their original definitions.

The use of jargon has the side-effect that it can exclude people who don't understand it. You certainly don't want to exclude your stakeholders on agile projects.

Therefore, pay attention to the language you use when talking to them. Try to avoid heavy use of agile terminology initially. Ease them into it gradually and be prepared to politely adjust your words in-line with what makes sense to them.

Respect Their Time

Stakeholders are busy with their own jobs. They are either voluntarily sharing some of their time to assist with your agile project, or they may have been *volunteered* by someone else.

It is unlikely that most stakeholders will be primarily focussed on your project. As such, coming along to participate in the numerous agile-specific activities and rituals has an opportunity cost. They will be giving up time on their other work.

Stakeholders are less likely to want to share their time if they don't perceive there is value in their participation. Additionally, if they are requested to join in on short notice, it is less likely to fit well with their own schedules.

If either of these situations unfolds, stakeholders just don't turn up to the activities they are invited to. I have seen it happen often. Review and feedback sessions become ghost towns.

The impact is your agile project stalls. Or worse, the project team must proceed with what they think the stakeholders want, absent of validation.

A way to support the need for time with your stakeholders is to give them sufficient notice. Let them know which activities would really benefit from their participation. Explain to your stakeholders why, when and for how long they will be needed.

Additionally, make it worth their time. Provide them value from the process. Show them that the time they are putting in has a major positive impact on the project.

You could even try to make their participation fun and enjoyable. Re-iterate the importance of their role. Show them that you like working with them.

Most importantly, respect their time. Get organised. Make sure you prepare well for your agile activities with the stakeholders. Start on time, keep things succinct and to the point.

Have goals each time stakeholders interact with your project and as soon as you achieve those goals, wrap the session up. Avoid running over-time.

Show Them Something Real

Have you ever heard the term *gunna*? As in, that person's a gunna. It's slang for going-to. It's a way of describing someone that is always talking about the things they are going to do but never actually does.

Agile projects that take too long to produce their first outputs are gunnas. Stakeholders hear all this talk about agile-this and agile-that, but they don't see anything tangible for ages. By the time something does turn up, the stakeholders have already made up their mind that this agile stuff is all talk and no action.

This issue is closely related to the last one about respecting stakeholders' time. If stakeholders are giving their time to participate in your project, but they are not seeing anything tangible being produced, how do you think they would feel?

They grow tired of waiting. They start questioning the value of the process. They begin to feel that their time is better spent on other things. And then, they gradually stop participating.

When this occurs, it has a flow-on impact to the rest of the project. The team members start feeling dejected and wondering if what they are doing is worthwhile. The project leaders start questioning the effectiveness of the project. A general malaise creeps into the entire project. It becomes a slog until, finally, something is produced.

Amazing things happen when real outputs get created on agile projects.

That sense of anticipation in the early stage of the project is fulfilled. People really appreciate and enjoy seeing their ideas and thoughts turned into real functioning outputs. It triggers those human emotions of achievement, satisfaction and pride.

Stakeholders really switch-on when the outputs start showing up. When they begin to review completed products and give their feedback, stakeholders start to experience what agile is all about. They start to notice the importance of their role once the first few outputs are placed into their hands and the team asks if this is what they need.

One way to support this is to structure your agile projects so that the team produces something small and tangible as soon as possible near the start of the project.

Avoid drifting off into deep analysis and design. Just get something out fast. Give your stakeholders something to play with. Trust me, the quicker you get something in the hands of your stakeholders, the better off the whole project is.

Once your stakeholders move out of the conceptual world and into real hands-on thinking, the tactile experience assists their ability to visualise what they truly need. They finally have something to evaluate and provide feedback on.

With real outputs, the ideas and suggestions that stakeholders come up with are a much closer fit to what they truly need. They can evaluate actual results as opposed to statements and models. This enables the project to get closer to achieving those target outcomes everyone is aiming for.

Display Commitment to Emerging Requirements and Priorities

I was recently sitting in as an observer to a planning session for a project. The project was creating a new office for an organisation I was consulting to.

It was a presentation from the building project managers back to the stakeholders. The stakeholders in this instance were a subset of staff in the organisation that would ultimately be working in the new office.

The building project managers were attempting to convince the stakeholders to dive in and accept a fully open and unallocated seating model. It was a trial of a new model that they were hoping to eventually rollout to the rest of the organisation over time.

The stakeholders were having none of it. Most of them had spent the majority of their careers working in either open-plan allocated workstations or private offices.

There was a significant disconnect between the building project managers and the stakeholders.

On the one hand, I could see that the building project managers were genuinely attempting to provide the best result for the staff based on latest industry practices.

The problem was, it was a massive change for the staff and they were very unsure how it would turn out.

The sticking point of the conversation was the fact that there appeared to be only one shot at designing and constructing the new office. Once it was complete, that was it for the next 15 to 20 years.

This was a showstopper for the staff. If there was no chance to modify and improve things with the new design, they would effectively be stuck with it.

Predictably, the staff responded by saying they would prefer to just stay with the same sort of design that they currently had. "Better the devil you know" is the saying.

I made a suggestion to the building project managers. Given the experimental nature of moving to this new style of design, perhaps they could commit to the emerging requirements and priorities of the staff in the new office.

That is, reassure the staff that there would be budget and time available in the years to come after the move to incrementally modify and improve the design of the new office. That way, the staff could dive in and try out this new fully open unallocated style with confidence that they could adjust it later if they needed to.

Of course, I understand it isn't easy to implement such an approach in reality as my casual suggestion would assume. There are physical constraints and challenges in such projects. Building and construction projects are costly endeavours that take time, effort and result in a lot of disruption. It isn't particularly easy to incrementally build and renovate an office multiple times.

However, what was being proposed in this instance would have a massive impact to hundreds of staff initially and potentially thousands later with the rollout of the design to the rest of the organisation. Getting it wrong would have lasting impacts.

You face a similar situation on your agile projects. Stakeholders are often uncertain about the outputs they are going to be stuck with at the end of the project. They are naturally sceptical and possibly uncomfortable initially.

One way you can assist in alleviating this scepticism and discomfort with your stakeholders is to provide commitment to their needs.

Show the stakeholders that they will be included throughout the entire project. Really involve them in the process of designing the outputs.

Give them assurance that things can evolve over time to meet their needs.

Lastly, and here is the crux of the matter, demonstrate that you will do what you say. As soon as you produce that first output, give the stakeholders the opportunity to provide feedback and then actually respond to that feedback. Make any reasonable adjustments to the outputs.

Prove to your stakeholders that they won't be ignored.

Once the stakeholders see that you have listened and responded to them, they will gain confidence in the agile-specific aspects of your project.

Make Feedback Easy

If you were a stakeholder on a project, which of the following two scenarios would you prefer? The scene is that the team has just produced a set of outputs. You have been invited to review their work and provide feedback.

Scenario A: Type your feedback into a document, providing details of all aspects of the outputs that worked well and not so well. Use a standard form with many fields to fill in. Hand the information back to the project team, wait for them to add their own notes, and then finally sign-off the documentation to prove everything was in order.

Scenario B: Visit the team and provide feedback in a conversation?

Most people would prefer Scenario B. The reason is that it is the easiest. It requires the least effort to complete.

On an agile project, you want to engage regularly with your stakeholders. If providing feedback is an onerous and difficult task that requires significant effort, stakeholders are less likely to want to participate.

This is one of the underlying ideas behind the Agile Manifesto. Especially the part that states "individuals and interactions over processes and tools".

To make stakeholder feedback easy there are few things you can do. Prefer face-to-face communication over formal documentation. Be available, and flexible, to their needs. Go and visit them, rather than making them come to you. If they want to write feedback down, let them do it using free-text.

A Mantra for Working with Stakeholders on Agile Projects

Everything we have spoken about so far in this chapter boils down into a few basic principles.

The answer to the question of how to work with stakeholders on agile projects is very simple.

All it takes is four actions, undertaken regularly.

Four simple actions can make a big difference to supporting stakeholders on agile projects:

Ask. Listen. Learn. Show.

Ask

Speak with your stakeholders regularly. Find out what outputs they need and how they want to work with them.

Ask your stakeholders if proposed solution ideas and designs match what they are thinking?

Seek feedback from your stakeholders in simple ways. Reduce the effort for them to work with you.

Listen

When your stakeholders provide feedback, listen closely to them. Record their explicit words.

Additionally, pay attention to their non-explicit communication. Notice their body language, actions, behaviour and reactions.

What are they saying? Does their behaviour match what they are saying? When they utilise the outputs, does it look like their needs are being met? Is there value generated? Are there any issues?

Learn

Stop and think about the feedback provided by your stakeholders. Try and understand things from their point of view. Show empathy toward your stakeholders.

Develop new ideas and solutions to meet your stakeholders' needs. Avoid doing the same old things that you are used to. There is no one-size-fits-all in agile projects.

Be aware that you may have unconscious bias. That is, your life experiences to-date may lead you to certain behaviours without you even realising it. Stop and ask yourself occasionally why you are proposing certain solutions. Is there another way?

Show

Produce outputs early and often. Demonstrate progress with actual tangible results.

When feedback is provided, prove that you have implemented their suggestions in updated versions of your outputs.

Build confidence.

With these four actions, your relationship and interactions with stakeholders on agile projects will be stronger and more effective.

This concludes Part II. You now have an insight into an often-overlooked aspect of successful projects. It's the people that matter. Agile projects work when people make progress together.

In the next part, I provide some hands-on tactics you can put into practice. It's time to look at the process.

PART III
PROCESS

Chapter 10
Let's Get this Project Started

Welcome to the part of the book where we roll up our sleeves and get our hands dirty. It's time to open the toolbox, have a bit of fun and make some noise!

The way this will work is that I will take you through a hypothetical example of managing a project. Remembering what I said back in Chapter 2, we'll run this example project using Scrum.

We'll go through the entire project from start to finish. Our example will begin at the very moment the project is conceived as an idea in someone's head. It will end with the popping of a cork sometime later once those target outcomes have been achieved and the benefits are flowing in.

Along the way, I'll show you everything I do throughout the entire lifecycle of an agile project. I'm pulling back the curtains here and letting you see everything behind the scenes.

I want to give you more than a book that tells you what to do. I want to share the exact tactics that I have used to successfully manage my own projects. My goal is that the art of project management is made available to everyone across the planet, equally.

As part of this open approach, I have created a set of tools and templates you can download and use to follow-on. The wonderful thing is, after you have finished this book, you can use them in your own day-to-day projects.

There is lots of great material coming up, so let's get into it.

The Project

When I started thinking about what to use as a hypothetical example project, I decided to approach it as an evaluation task. What are the criteria relevant to selecting the project?

First, I wanted to be certain that you worked with a project that had nothing to do with software or the IT industry. After all, as I have stated quite a few times, agile is not just for software development projects.

If you, dear reader, work in the software/IT industry think of it as an opportunity to gain experience outside of your current profession. If you work in a different profession, I hope this provides you with material that is non-IT related. It does seem that there is relatively more IT-specific material relating to agile out there in the world already. Let's try and nudge the balance in a more general direction.

Next, I wanted the project to be significant enough that it covered a range of features that you would find in a real-world project. However, I didn't want it to be too complicated or long-running. The intent of the book is to get you working with agile, not to have you stuck working through the details of a multi-year mega-project.

Third, I wanted the example to be the type of project you might be familiar with already or that you might be able to apply in your own business one day. The idea is that you can easily follow-along, with interest. As much as I'd love to use an example of building a rocket to fly us to Mars, it's probably not something I would ever find myself using in reality.

By the way, if you, dear reader, actually are a rocket scientist and you are using agile on your projects, please get in touch with us! I suspect that a lot of people in the *Don't Spook the Herd!* community would love to hear about it.

Without further ado, I introduce the Agile Conference project!

We're going to organise and run an international conference. The project will bring together the world's leading voices and practitioners of agile for a multi-day event aimed at advancing the profile of the industry, sharing new ideas and creating unique networking opportunities.

Let's double-check this meets the evaluation criteria. Running a conference event is non-IT specific, so that passes the first check.

We will ensure our conference has a big budget, long timeframe and a jam-packed activity list so we can consider it *sufficiently* significant.

Lastly, even if you are not in the conference hosting industry, you could visualise how such a project could be applied in your own business. Think of the elements of a conference that could be applicable to any business. For example, it is likely that most businesses would need to organise a workshop or get-together of some sort. If you use your imagination, there will be elements of this example project that you could take-away and use in your own career at some point.

Ok my friend, grab your kit and let's get cracking. It's time to put agile project management into practice.

As you read through the following chapters, bear in mind the purpose is to work on a hypothetical example. Occasionally, you might find yourself getting immersed in the experience. Other times you might find yourself wanting to skim over the details. The choice is yours. Do what feels best for you and your own preferred style.

My main goal is that this book provides you with a real hands-on guide to running an agile project.

Chapter 11
The First Few Days

Picture this, it's 10am and you have just finished wrapping up a decent chunk of work. You step away from your desk and grab a nice warm cuppa' for a well-deserved break.

You've been hard at that last piece of work for several months. Time to reflect on a job well done.

No sooner do you get to take the first sip when your CEO wanders past and asks if you've got a minute. Since you're always happy for a chat with the Chief, you head on into their office.

"I hear you had some recent success with that last project you were working on?" the CEO enquires. Smiling, you acknowledge that you enjoyed it and can't wait for the next one. "Great!" exclaims the Chief, "Well, we've got a new project coming up that I think you'd love just as much. Let me tell you about it".

At this point, the CEO and yourself discuss the upcoming conference project. You are offered the opportunity to manage the project, which you gladly accept.

The Chief explains that there is a fixed budget and rough target date for the event. The initial key stakeholders in your organisation, *Project-Corp*, are identified and the Chief appoints a senior executive to be your designated go-to person.

It is also mentioned that your organisation, Project-Corp, has a large mailing list of potential attendees and presenters for the conference.

Being a well-known world-leader in project management education and project consultation services, it is likely you will have no trouble getting the word out to tens of thousands of people, many of whom are likely to want to come along. You thank the Chief for the opportunity and confirm that you'll get right onto it.

Walking back up to your desk, you take the last sip from your cup and imagine what's to come. You visualise a huge conference centre jam packed with interesting people from all over the globe. You see auditoriums full of eager participants sharing ideas and making new friends. You also see some great post-event dinners and drinks!

"Well then", you think to yourself with a smile, "where do I start"?

Could This Work as an Agile Project?

You've been looking for an opportunity recently to try out the agile project management practices you have been learning about. "Could this be it?" you wonder.

Let's look at the characteristics of this project and compare them to the criteria we identified back in Chapter 3. To do this, we use a handy little *Agile Decision Checklist* I created.

As a special treat for anyone reading this book, I have made available a set of agile project management tools and templates. They are yours, for free, to access anytime you like. I host this companion material on my website, https://miller.productions/.

Download a copy of the Agile Decision Checklist from https://miller.productions/agile-toolbox/

Now that we are about to use tools that I am providing for you, please read the full disclaimer in the copyright page at the start of this book. Reasonable common sense is in order. My goal is to support you on your agile journey, but you should make your own decisions about when and how you use these tools for real.

I use this decision checklist all the time on projects that I suspect might be candidates for running with agile project management.

The way I use it is to think about my new project in terms of each criteria in the sheet. Once I reach the end, the more *yes* answers I have, the more likely the project may be suitable for agile.

As I noted earlier, not every project is suitable to running with agile project management. Even though the decision checklist may *indicate* suitability, that doesn't mean it is true.

If there are any items left unanswered, or if there are any burning questions in the back of my head, I investigate them in more depth.

Let's now go through this decision checklist with our example conference project.

Is the Project Unique?

In our example, running conferences is not something this organisation does very often. In fact, the conference project is the first of its kind for you and your colleagues. You have run smaller events before but never anything on this scale.

While there have been thousands, if not millions, of conferences run previously around the world, you and your organisation have never done it before. Therefore, yes, this project is unique.

Are the Requirements Uncertain?

So far, the conference is nothing more than an idea from the mind of the CEO. We have little information apart from a rough idea of when it is to take place and where to initially get ideas for attendees and presenters.

You anticipate that there will be several key stakeholders on this project, each with their own opinions and ideas about what the requirements should be. However, nothing is confirmed yet and you anticipate that decisions about requirements will emerge.

It would be fair to say, yes, the requirements are quite uncertain at this point.

Can the Plans Evolve Gradually?

There would be a few items in the conference project that would need to be known up-front. For example, you would want to know when and where you were hosting the conference. That is, roughly what dates would you run the conference and in which city?

For other aspects of the project, however, it is possible that you could determine the details gradually in the lead up.

In fact, if you try to lock in too many of the aspects of the conference early on, there is a risk that you could over-commit. The entire venture could become massively unsuccessful across a wide range of criteria.

For example, the biggest risk is the cost of the event. Without knowing ticket sales in advance, you could end up booking too much venue space, equipment, materials and staff. Thus, the conference could become a major loss-making activity for your organisation.

The conclusion is: yes, the plans could evolve gradually. In fact, you would probably want them to on this type of project.

Is the Cost of Changing Direction Acceptable?

This is a tricky question to answer for the conference project. It depends on what sort of deals you could make with the various suppliers for the conference.

For example, many companies in the business of supplying to conferences require up front deposits. For example, a conference hall requires a deposit up-front to ensure availability. Equipment such as audio-visual gear also needs to be booked in advance.

If you decide later to change venues or use different equipment suppliers, you could lose your deposits. Additionally, the longer you wait to make your bookings, the less chance you have of availability.

Like I said, though, none of this is certain and will depend on the situation.

For the sake of the example here in this book, let's assume that we are in a big city with plenty of different conference venues and suppliers to choose from. It might be easy to switch direction for some things, while others might involve the loss of deposits.

Thus, we shall conclude that there is a moderate cost of changing direction for this project. The question is, however, is this cost acceptable? We would need to run it past the Chief to get a feel for it. The conclusion is, we are uncertain if the cost of changing direction is acceptable for this project.

Are the Project People Flexible with Change?

In this hypothetical example, let's assume that yes, the people involved with the project are flexible. This is an organisation that specialises in project management education and consulting. The people are familiar and comfortable with modifying direction as required.

We can safely make reasonable adjustments to project targets and output designs and the project people will be OK with it.

Will the Key Stakeholders Participate?

This project is a strategic activity, being driven by the CEO. We could safely believe that the key stakeholders from our organisation would be committed to participating.

As for people outside the organisation, such as suppliers and representative attendees, that depends. If there is money to be made, it is more than likely that suppliers will participate. For our representative attendees, we could put in place some form of incentive, cash payments or in-kind gifts, for anyone willing to commit their time to assisting with our project activities.

In this example, let us say that it is still uncertain whether the key stakeholders will participate.

Are the Leaders Comfortable with Gradual Information?

The CEO handed this project to you with nothing more than a conversation and description of a few minor details. Given this, let's say that the leaders for this project will be quite OK with gradual information.

Would Stakeholders be OK with Possibly Missing Out on Some Outputs?

If the project were to be run using agile, would the stakeholders be OK with having their desired outputs prioritised? Would they also then be OK with the project potentially not delivering every single output on their list.

That is, as we now know, in an agile project the lowest priority outputs get worked on last. Hence, there is a chance that some items won't get completed. It depends on capacity and throughput of the team and other related constraints.

In this hypothetical example of the conference project, we simply don't have enough information to answer this question yet. Once we get to know the stakeholders further, the answer may emerge. But for now, it will need to remain open.

Will There Be a Relatively Small Project Team?

To answer this question, let's say you contact one of your colleagues outside of your organisation. This old friend of yours previously worked in conference event management and can help advise you.

You find out that some conference event projects can be run with a core team of around seven or so people. Supporting this team would be other additional suppliers and support staff. However, the suppliers and support staff would not be involved day-to-day.

The answer to this question is deliberately subjective. What is *relatively small*? I'll leave the answer up to you and your own experience. There are numerous research papers available on the subject of team sizes, but ultimately, it comes down to what you and your team are most comfortable with.

At some point, adding more people to a project team makes things complicated. Communication overheads creep in, things take longer, information doesn't get through to everyone, confusion sets in. As to what size team this starts happening with depends on many factors.

These factors include what industry your project is in, what type of outputs it is producing, what sort of environment and tools you are working with and how experienced the team members are.

For the sake of the example, let's conclude yes, the core project team size is relatively small.

Is Incremental Delivery OK?

Would it be OK if the outputs of the conference were delivered gradually? Let's say yes to this one.

The CEO has given us sufficient time for the outputs to be produced. Given it is a conference, there are many outputs we can work on ahead of time, before the actual conference event itself.

In fact, the more outputs we can produce gradually before the main event, the better idea our stakeholders will have regarding suitability.

Our stakeholders will be able to see the plans, visualise the conference rooms and stages, review draft content, etc. Yes, incremental delivery will be OK on the conference project.

Would Imperfect Solutions be OK?

In the lead-up to the main event, i.e. any-time before the date of the actual conference, imperfect solutions would be OK. In most productions that involve an audience we call these early work packages reviews, dry-runs, rehearsals, etc.

For a production event with a large audience, it probably wouldn't be ideal to have imperfect solutions on the day. You want everything to run smoothly so the attendees have the best experience possible.

However, you can practice this as many times as you like beforehand.

You might appreciate getting things wrong during the preparation stage so that you catch all the bugs and iron out the kinks so-to-speak.

With this in mind, yes, imperfect solutions would be OK.

There is a caveat on this, however. Imperfect solutions are OK in the lead up, but we want perfect solutions at the live event.

Thinking This Through

At this point, we have answered the questions on our Agile Decision Checklist. What does it tell us? In our example, the results are:

- Yes: 8
- Uncertain: 3
- No: 0

Ok, so we have mostly yes answers and a few uncertainties. What I would do at this point is to arrange conversations with the CEO and any other relevant stakeholders to learn more about those uncertain items. I would also check my assumptions to ensure my answers matched what others were thinking.

In our example, let's say you successfully complete this and you are satisfied with the answers. You decide that the conference will be suitable for running with agile project management.

To help support this decision, you do a little more research to see if there are examples available of other organisations using agile with this type of project. You find that there is and you also contact a few organisations to see what their experience was like.

You have a match. You decide that you will run the conference using agile project management.

Which Flavour of Agile?

The next step in the process is to decide which flavour of agile to use. The obvious thing to note right now is that the conference is not an IT project. Therefore, we can exclude a few sub-genres from the list we went through back in Chapter 2.

You ponder to yourself, "would this project be a candidate for working with Scrum"?

Is there are fixed timeframe for the project? Yes. The Chief has given us a preferred date range for the event to take place.

Could we build up the project outputs via a series of iterations? Yes. The outputs for a project such as a conference could be delivered in blocks or packages. In the lead up, the project would benefit from delivering outputs as a series of packages.

Would the project benefit from prioritising the work and targeting specific outputs each iteration? Yes. Doing so would enable the project team to focus on specific areas each iteration.

When the event goes live, all outputs need to work well and integrate smoothly. Seeing the work come together will help the project stakeholders gradually build up their vision of the final event and adjust their ideas along the way.

Would the project team and stakeholders benefit from checking-in on progress at regular intervals using demonstrations of completed outputs? Yes. A pause to take-stock at regular intervals would enable the team to check the quality of work completed. We would much prefer to pick up issues in the preparation stage rather than discovering problems during the live event.

Will the key stakeholders of the project be highly engaged and committed? Possibly. This one we're not certain about yet, so we will need to come back to this.

Are the team members comfortable working with formalised processes and management frameworks? Once again, possibly. We'd need to check with them and keep an eye on it.

You also do a similar evaluation around the aspects of Kanban but conclude that it would be less of a good fit with this project. There are a few reasons around this. One of the main reasons is that everyone involved in the project appears to have little previous experience with conference projects.

Just setting the team to work and having them keep grinding it out until the date of the main event might not work so well. You need a means to check in with them regularly to ensure progress toward the targeted combined set of outputs.

With Kanban, the team would be working on outputs, one after the other. This is great for outputs that are quite similar and where each one could be produced within a few days. The team members could focus on continuously delivering and gradually improving their output rate.

However, the conference project outputs are varied and likely to take a bit longer than one or two days each to complete. The aim of the conference project is not to keep delivering fast, but rather to deliver the most appropriate and highest value set of outputs.

Also, Kanban would enable the priorities to be switched regularly so the team is focused on responding to short-term stakeholder needs.

On the conference project, the team needs to focus more on the overall set of outputs as a complete package, not just a series of isolated jobs. Allowing priorities to fluctuate regularly might distract the team members from staying on track.

At this point, you decide that Scrum would be a good fit for the conference project. The next step is to get some of the high-level facts of the project understood and written down. It's time to begin sharing the initial details with others

Chapter 12
State Your Intent

It is now a day later and you are working from your home office.

You assume that others back at the main office must have also had similar conversations with the CEO about the conference project, since you are starting to hear from them.

You have received a couple of messages from your colleagues where they mention they would like to get together with you to talk about the upcoming job.

Wanting to stay on the front foot, you decide to put together an information pack that summarises the plan for the project. That way, you can take it with you to each of the conversations you are about to have with your colleagues.

A short document, i.e. a *one-pager*, that gives an overview of the project is a great way to help bring everyone together right from the start. It lets people know that you are actively leading and that you have a plan. It also provides the initial details of the project in a nice succinct format.

Remember that phrase in the Agile Manifesto? Working [outputs] over comprehensive documentation. We're about to apply it now.

In this chapter, I discuss a brief history of project documentation and then show you how to use an often underestimated, but powerful, agile project tool: the *Intent Statement.* What looks to be a humble sheet of paper, turns out to be a guiding standard for project documentation.

The Dog Days of Documentation

Back in the old days of managing projects, you could literally have piles and piles of paperwork sitting on your desk at any one time. This paperwork recorded every small detail about the project. Every time there was a revision to a specific document, you would end up with an entirely new updated copy of that document.

Heaven help the poor souls that used to spend their days writing and producing such copious documentation. It used to be that project managers would spend the bulk of their time writing content and keeping it updated.

I'm sure it made the leaders of those projects feel quite important. The people of that era must have developed some perverse relationship between the quantity of documentation produced and the merit of a project.

Historically, this idea that documents were so important to a project harks back to the days of engineering and construction.

In a building project, for example, there would be numerous crafts-people and trades-professionals working on the project over time. The instructions for the project were conveyed through drawings and specifications.

The details of what to build and how to build it needed to be precise and detailed. If it wasn't, there was a risk that different people on the project might have their own different interpretations. This could be a disaster. Just imagine walls not lining up, roofs with gaps, etc.

Additionally, project documentation served the dual purpose of both contract and evidence. People running the project would write down every little thing they could think of including the specifics of time, cost and scope. This would be provided to suppliers who then completed the work.

As is inevitably the case, things occasionally turned out differently than what the documentation specified. Perhaps the suppliers were not able to use a particular material as planned. Or maybe parts of the job turned out to be more complex than originally thought and thus took longer and cost more than expected. The owners of the project would be able to refer to the sections in the documentation that their suppliers didn't complete as agreed.

In a similar way, suppliers would produce their own documentation that confirmed exactly what their clients asked for. If there was ever a scope change, the documentation would be updated. This way, suppliers could refer to their own documentation if the clients questioned the final scope, time or cost.

Over time, as the project management profession began to emerge, this approach to heavy documentation was transferred into other types of projects and industries. After all, it was what worked at the time. It became the norm for projects to produce documentation. Whether it was to ensure precision in the outputs produced, or as an instrument of indemnification and legal defence, documentation was a major part of projects.

There are cases to this day where it is still valid and fair for a project to use detailed documentation. We will always have the types of projects that require the precision of drawings and specifications. I'm pretty sure the lawyers of the world won't be going anywhere either, it's too lucrative a profession for them. Thus, we are probably going to keep seeing project documentation as contract and evidence. However, there is a comforting breeze on the way.

The Dog Days are Over

If you have ever used the terms *executive summary* or *abstract* before, you would understand the concept of condensing information into an abridged form. The idea is to get straight to the point and tell the reader only what they need to know for the given context.

Most of the executives and chiefs in the world are flat out busy most of their lives. They normally have several responsibilities on the go at any one time. Their roles require them to use their skills and experience to make numerous decisions every single day. Many would prefer a quick conversation rather than a lengthy stack of written paperwork to support their work.

Documentation in agile project management is a lot like using only executive summaries.

Everyone on the project is encouraged to use short succinct communication to analyse requirements, convey ideas and share information. The preferred method of communicating is via conversations between people.

In agile projects, the guiding principles are that everyone focuses more of their effort and attention on individuals and interactions, working [outputs], customer collaboration and responding to change. While still having a place, less value is placed on processes & tools, comprehensive documentation, contract negotiation, and following a plan.

My own interpretation of this is…

If it must be written, keep it simple.

The first few times some folks hear about agile, their initial reaction is to think of it as the approach where nobody writes anything down.

My own interpretation of the relationship between documentation and agile project management is that the agile approach improves the value of documentation.

Rather than ending up with reams upon reams of written material, as well as time wasted producing and updating that material, agile projects only write something down when it is necessary. The content that is written also gets straight to the point with little fuss.

I am going to show how I set the scene on agile projects to encourage such minimal documentation.

The Intent Statement

One of the first things you need to do when starting up any project is to communicate with people. You must speak with potential stakeholders and team members to let them know that the organisation has decided to work on this activity.

As quickly as possible, your goal is to create a shared understanding among anyone that needs to know. At this early stage, the basic information includes what the project is, why it is taking place, who will be working on it, when and where it is happening and how it is likely to be undertaken.

When you speak to people, you also want to leave them something behind, so they can remind themselves of the details later. You want them referring to the same material, so the message and understanding is consistent for everyone.

The tool of choice to meet these early communication goals is the Intent Statement. A one-page summary of the key details of the project. Let's create one now for our example conference project.

Download a copy of the Agile Intent Statement from https://miller.productions/agile-toolbox/

I use this tool on every agile project I run. The intent statement serves two purposes. The first is that it provides the main details for the project in a succinct form. Second, it lets everyone on the project know that the documentation produced will be minimal. In other words, it sets the expectation and style for others to follow.

The remaining material in this chapter closely follows the actual intent statement from my site. I encourage you to download it now so the next few sections are easier to follow and understand.

As you can see, our intent statement is divided into a few parts. There is the title at the top and a three-column body containing various sections that describe the who, what, why, etc. of the project.

Title and Pitch

Give the project a name. Underneath that, write a short sentence that sums up the main outputs and benefits the project is aiming for. Don't overthink it or try and fit too much here. Consider it your elevator pitch. The one sentence you could use repeatedly as an introduction when telling people about your project.

In our case, let's put something like "Host a conference to enhance knowledge of agile project management for the Project-Corp community". This is just a hypothetical example, so feel free to come up with your own title and pitch. Practice keeping it short and sweet.

Background

This is the first section of descriptive content in the intent statement. It is a short paragraph or a few dot points that summarises the history of what led to this project.

For example, what is the current state of the world, in the context of this project? Why do we need or want this project to go-ahead?

Benefits

Here we list out the benefits that the project is aiming for. The key to long-term project success is knowing what your goals are.

When we express the goals of the project in terms of benefits, we can then work backwards to ensure that each activity in the project along the way is designed with those benefits in mind.

For our conference project, we eluded to the benefits in our pitch above. Elaborating on them a little further, we could propose the project is aiming for one main benefit:

- Increased sales of Project-Corp's products and services;

Note, later in the project you will want to go deeper into the analysis of the benefits the project is aiming for. For example, you might wish to apply the S.M.A.R.T.[xiv] approach on them.

S.M.A.R.T. is an acronym used to describe an approach to setting goals and objectives. The approach encourages goals and objectives to be specific, measurable, assignable, realistic and time-related.

However, for now we're simply going for rapid sharing of thoughts and ideas to help communicate with others about the upcoming project.

Target Outcomes

These are the results that are attained when the project's outputs are utilised. That is, an increase or decrease in the specific units or measures that are drivers of the benefits listed previously.

In our conference project, we could target the following, for example:

- increased knowledge of agile project management for the Project-Corp community;
- increased demand in the market for agile project management education and project consulting services;
- increased awareness of the Project-Corp brand in the markets for project management education and project consulting services.

Outputs

Here we list the high-level outputs of the project. In other words, what are the main items that the project will produce? In the conference project, our initial assumption is as follows.

- Two-day agile project management conference.
- Marketing and promotional material.
- Attendee market research report.

Once again, bear in mind your project is likely only a few days old as an idea at this stage. The composition and details of your outputs will evolve over time.

For now, a simple starting point is all we are looking for. That is, something to get the conversations started so that details will be forthcoming in the near future.

Participants

In this section, we list the names and groups of people involved in the project. I like to follow the categories specified in Chapter 4 listing the names of people in individual roles, such as the funder, and leaving groups, such as the Board and project team, as general titles. In our conference project, the participants are:

- Funder: Ms Penelope Persons, CEO.
- Owner: Mr Joe Citizen, Executive Manager.
- Project Manager: Mr Dan Miller (or yourself).
- Board: Key representatives from Project-Corp, conference speakers, and associated suppliers.

- Customers: Interested participants from the Project-Corp client list and the global project management community.
- Project Team: members from Project-Corp, supported by external consultants and contractors if required.

Location

This one is fairly self-explanatory. We are going to let people know, through the intent statement, where the applicable locations are for the project.

This could include where the project team will be located most of the time when doing the work to create the outputs. There could also be an additional location if the final outputs will be used somewhere else away from where they were created.

For our conference project, let's assume we have a few locations:

- Most of the people in the project team will be working together in Canberra, Australia.
- Some team members will be working remotely from Sydney, Melbourne, Perth and Brisbane.
- Occasional excursions will take place to observe and learn from other well-known conferences.
- The conference itself will be held in Sydney, Australia.

Timeframe

For this section, simply list the key dates for the project. This will at least include the proposed start and end timeframe for the project. It could also include any specific milestone dates.

In our example, the CEO has previously mentioned a rough target date for when the conference should be held. Let's assume the following:

- Start: February 2018.
- Live Event: late February 2019.
- End: March 2019.

Do you notice that I have only listed the months and not the days for the start and end of the project? This is deliberate and is designed to manage the expectations of the project stakeholders.

On any project, there are usually several unknowns, risks and issues that can affect calendar dates. Even the most experienced project manager will be impacted by things outside of their control.

The moment you start listing explicit dates, people expect them to be set in stone. When the actual dates for key milestones on a project differ from estimated dates, some stakeholders tend to experience concern. Missing a date can be perceived as a failure. After all, society conditions us to pay attention to dates. We start and end our standard business days at set times. We like to be on time for meetings. Turning up late is frowned upon.

However, in the early stages of a project, before you have had a chance to plan things out with your key stakeholders, it is advisable to keep your timeframes flexible. As the project progresses, you can begin to tighten up those dates to be more specific, but still take care when doing this.

Of course, for something like a public event, our agile conference for example, there absolutely must be a day and time set. We are unable to bend the laws of physics and thus the conference must happen on a specific date. We can leave it as late February 2019 for now in the early stages of the project. We can then tighten it up to a specific set of days once we confirm a few things such as availability of people, suppliers and venues.

Approach

In this section, you briefly describe how the main outputs for the project are going to be produced. The information here should be kept at a very high level. Here is something we could say for our conference:

- Event planning and co-ordination to be delivered by the core team from project-corp.
- Event venue and associated logistics by external suppliers.
- Project managed using Scrum. Outputs delivered through iterations with review and feedback opportunities.

Spread the Word

Now that you have produced the first version of your intent statement, you visit each key stakeholder for the conference project and provide them a copy.

What they receive is a consistent introduction from you and a shared understanding of what is coming up.

In the next chapter, we discuss the activities associated with bringing the project people together. The legwork and efforts you have put in to create and share your intent statement will serve you well. People will be ready to hear from you.

Chapter 13
Welcome to the Crew!

In any group activity that involves people working together to produce something, those people need to know their roles and responsibilities.

If you were creating a sporting team, you wouldn't have much success by pointing at the field or arena and telling everyone just to get out there and play. Similarly, telling a bunch of actors and production crew members to make a movie without assigning roles and jobs would be an absolute disaster.

The same goes for projects. You can't just tell a bunch of people that they are working on the project and then, voila, hope they will figure everything out.

In this chapter, I show you the techniques I use to welcome people to an agile project and get them productive as quickly as possible.

The first step in the process is being aware of the various roles that need to be performed. To explain to someone what they are expected to do, you'll want to understand the details yourself.

Next, identify who will be assigned to each role. This could be as simple as nominating staff, or it might require a competitive process.

Last, it's time to approach people individually and confirm their appointment to each role. The goal is to ensure they understand their responsibilities and that they are committed to giving it their best.

The Selection Process

Before you embark on the activity of forming your project crew, consider the importance of picking the right people.

Be careful of making assumptions at this point. It is important to speak to each individual, preferably face-to-face, and confirm a couple of things.

First, would they like to take on the role you are proposing? Second, would they be a good fit for the project? Treat such discussions as an interview, so-to-speak, to get a good feel of the value each person would bring.

Depending on the availability of people from within your organisation, you may have to go out to market to bring in new staff for the project. You may need to hire people to work on short-term temporary contracts, or your organisation may have the resources to hire new permanent staff. Either way, take care.

You are looking for people that are strong in a few areas:

1. Try to get a good fit to the agile characteristics we described in the various chapters of PART I of this book. That is, look for people that would work well with agile projects.
2. Go for people with a demonstrated and verified delivery focus. In other words, don't just rely on their statements in an interview. Get them to show you examples of their work and follow-up with reference checks. Agile projects require people that can make decisions fast and produce.
3. Look for problem solving ability. Projects are unique and will inevitably involve moments where people need to solve interesting problems to produce and deliver outputs.
4. Communication and relationship skills are essential. Remember, people are the key to successful projects.

Generic Project Roles

Back in Chapter 4 we introduced different roles and groups for people involved in a project. Your first step in forming the project crew is to identify who would be in each of these roles and groups.

We had an initial run at this back in the previous chapter. What you would do from here is to contact each of the people, or representatives of each group, and confirm whether they would be a good fit for the assignment.

Essentially, what you are doing is a first-pass stakeholder analysis. It is worth recording your analysis in an appropriate format for later reference. Once again, I would refer you to the work of Zwikael and Smyrk for ideas here. They provide a great framework for stakeholder analysis in their book[xvii].

Agile Scrum Roles

There are three roles recommended in Scrum projects, namely:

- Product owner;
- Development team;
- Scrum master.

While the names used by Scrum sound specific, they are often synonyms of existing general project management terms and phrases that you might ordinarily be familiar with.

The product owner is equivalent to the project owner role we described in Chapter 4. This is the person that represents the views of the key stakeholders in respect of the outputs being produced. They are also responsible for helping define the priority order of the outputs in the project backlog.

The development team is the group we previously labelled as the project team. This is the core group that works closely together daily to produce the outputs.

Scrum master is a fancy title given to someone on the project that will help keep things moving as well as guiding/coaching others regarding the Scrum processes. Think of them as a co-ordinator.

Many organisations and agile professionals often recommend that the role of scrum master is assigned to a single person. I agree with this. A good scrum master can make a big difference to the efficiency and productivity of a Scrum project. But this isn't the only way. There are also scenarios where you might be able to share the scrum master role around.

On several projects I have managed, we have rotated the scrum master role through each member of the development team. One person takes on the role for the duration of a sprint and then hands it over to someone else for the next sprint, and so on.

The aim of this rotation is to provide career development opportunities for staff members. It also helps reduce the administrative burden of the role from overloading a single person for extended periods of time.

There are, however, a few caveats to rotating the scrum master role. It doesn't work in all scenarios. It can backfire if you're not careful.

First, rotation works well with advanced/highly-productive teams but not so well with less experienced teams.

Some staff with advanced skills tend to enjoy taking on the scrum master role for the new challenge and experience it provides. Whereas, less experienced personnel have enough on their plate already and sometimes see the additional responsibility as an overhead they are not ready for.

The second caveat to rotating the scrum master role is willingness.

In some cases, project team members genuinely don't want the distraction. They prefer the freedom to focus on producing outputs. That is perfectly acceptable. What we do to mitigate this issue, is simply ask for volunteers each rotation. Anyone that wants to join in, does so willingly.

Show People How They Fit

Once you have determined which roles have been assigned and which roles or groups are yet to be determined, it is time to get the word out to everyone.

One way to do this is to use a Project Organisation Chart. If that sounds familiar, it's because it is just like any other organisation chart you may have seen before, simply focused on the context of the project.

An organisation chart is a diagram that shows the structure of a group and the relationships between the people/roles within the organisation.

Download a copy of the Project Organisation Chart from https://miller.productions/agile-toolbox/

Using this tool is easy. All you need to do is check that you have the correct roles represented and then fill in the names as you go. The composition of the two groups, project board and project team, will vary between projects so you will need to tailor these each time.

You will initially have a few blank entries but don't be concerned about this, the details will emerge over time. In other words, you will have identified the initial roles for your project but won't yet know which people are going to be assigned to every single one of those roles. Some may be unassigned for the first few days or weeks of the project. The point is to get the diagram out to people so they can start thinking of themselves in terms of their relationship with others on the project.

Pay Attention to the Board

Pay close attention to the project board section. The people and titles included here can vary based on the feelings of the leaders in your own organisation.

It is possible for the number of people represented in the project board to grow too big if you let it. Funnily enough, managers like to see their name included in diagrams like this. You will often have no shortage of people wanting to be included in the early stage. However, as the project progresses you find that too many people in this group can make the administration and decision making on the project unwieldly.

Talk to the Team

The largest group represented in your chart will be the project team. This is also the group that will be spending the most time together for the project. It is important when representing someone that you use the correct words and that all is in order from their point of view.

I like to share a draft copy of my organisation chart out with the team first before sending it out to the wider world. That way, the team gets to confirm that I have described their professional skills and specialities using terminology that they are comfortable with.

It also gives everyone an opportunity to visualise who else they will be working with and to speak up about any potential issues they might foresee. You occasionally need to be aware of quirks between people. That's just the nature of being human I guess.

Keep it Updated

Over time, people are going to come and go on your project. Whenever this happens, be sure to make an updated version of your organisation chart to reflect the change. Then once you're ready, share it out with the project stakeholders. Doing this reinforces the roles for people on the project and helps strengthen relationships. When people see themselves as part of a group, that evolutionary human instinct of team bonding kicks in.

We're beginning to see more details of the project unfolding now through our early analysis and planning. From here, we have enough information to take a closer look at the finances for the project.

Chapter 14
Show Me the Money

At this point you are ready to refine your budget estimates. The conversation you had with the Chief earlier gave you a bit of an idea about the available budget. Now that you have confirmed the initial crew, you can refine your cost calculations.

If you are wondering why we are starting to get into the details of budgets and costs in a book about agile project management, consider this. Projects have limited funds to run with. People and suppliers need to be paid for their efforts and materials. Knowing how much money is available to pay for the people on the project, tells you how long they will be available for.

Agile projects are very much centred around the concept of time. There is a focus on how much output the team can produce given a fixed duration. Establishing the budget determines that available duration. Thus, the budget is very much a constraining force in an agile project.

This book does not provide a detailed guide to project budgeting and cost management. That is an entirely separate artform itself. Instead, I provide a summary in this chapter of the main aspects of budgeting and cost management that you should be aware of.

I certainly encourage you to go beyond *Don't Spook the Herd!* For further understanding of this topic. Being financially literate and having sufficient skills in budgeting and cost management is a precondition for successful project management.

Labour Costs

On a project such as the agile conference, the people involved are going to make up a large proportion of the overall costs. This type of expense is otherwise known as labour cost.

The general idea is that you identify which people are going to be working on your project and for each person, determine:

- How often will they be working on your project? Will they be working consistently from a given start day to some end day? Or, will they work sporadically? That is, will they drop in and out of the project for varying blocks of time?
- For each block of time someone is working on the project, what percentage of effort will they be allocated? The easiest way to estimate this is to consider it per week. If someone is allocated to the project full-time, i.e. 5 days per week, then that is 100%. If they are only working for 2 days per week, then that is 40%.
- For each time block, multiply the number of weeks by the percentage allocation. Add these up and you get an estimated number of *effort days* that this person is going to be working on the project.
- Multiply effort days by that person's daily charge rate and this gives you the estimated labour cost for that person.

Do this for each person on the project, add them up and you have an estimate for the labour costs of the project.

Things to watch out for with labour costs are overheads or on-costs. In some organisations, staff can have a nominal cost per day but there may be additional costs associated with each person. These could include sickness allowances, holiday allowances, pension or superannuation payments, insurances, etc. Some organisations might also allocate the cost of office space or electricity and other such expenses on a per-employee basis.

It depends on each organisation. As a project manager, you should become familiar with how your organisation calculates labour costs.

Non-Labour Costs

Having confirmed the labour cost, you can determine how much is leftover for the other non-labour aspects of the project. Non-labour costs include items such as:

- equipment;
- supplies;
- materials;
- travel;
- training;
- catering;
- facilities;
- etc.

Given the infinite types of projects out there, it would be impossible to provide specific instructions in this book on calculating non-labour costs. Instead, look at it from a generic point of view:

- List each of the non-labour items you expect to need. For our conference example, this might include venue hire, equipment, marketing and promotion, etc.
- Assign a cost estimate to each non-labour item.
- Distinguish between variable and fixed costs where you can. This helps later if you need to make any adjustments.

Budget as Master Constraint

Together, the labour and non-labour costs of the project provide you with an estimate of your overall budget. This budget, my friends, is usually *the* master constraint on an agile project.

You will notice this constraint regularly as the project progresses. It mainly comes up during output prioritisation and solution designs.

With a fixed budget, you have a fixed timeframe within which the project can run. With a fixed timeframe, the project can only deliver a limited amount of outputs.

Agile project management aims to deliver the highest value outputs to the project stakeholders within the given constraint. Agile helps support the hard decisions that must be made about what will and won't be delivered. Such a constraint helps set expectations with all stakeholders.

This constraint means that some outputs will be completed, while others won't. It also means that the team will need to focus on delivering just-enough for each output and nothing more.

As you can see, what gets delivered in an agile project ultimately comes back to how much money is available.

Monitor Your Budget

As your project progresses, you will want to monitor your actual expenditure against the planned expenditure. This enables you to determine if you are tracking over, under or on budget.

It will help if you divide your overall budget down into smaller timeframes. For example, on a longer running project you might divide the budget down into monthly amounts. Similarly, for smaller durations, you could divide it down into weekly amounts.

If you go over budget, you may find the project needs to be re-analysed. Your project owner and funder have a few choices available which basically amount to adjusting the remaining scope, time or costs. Each treatment is going to impact the overall results.

If such a situation occurs, it flows through to your team's ability to focus and deliver. It is basically a negative spiral or feedback loop.

Pay close attention to the budget. It is the lifeline of an agile project.

Now that the details of finances are being taken care of, we can move on to the schedule.

Chapter 15
Build a Time Machine

We have a budget, we have a rough timeframe and we know who our initial team members are. With this, we have enough information to create an initial schedule for the project.

In this chapter, I take you through the steps of creating, maintaining and tracking a schedule for a Scrum project.

Gaze into the Future

Did you know that good project managers have super powers? They do, I swear! According to stakeholders using the old traditional styles of project management, it is possible to create a schedule that predicts the future.

Apparently, it is plausible to inform everyone about exactly what outputs are going to be built on a project, how long each will take and when they will be ready. Can you believe it!

Hmm, well, I'm sorry to rain on their parade and spoil the story but the truth is, such foresight isn't always achievable. For a lot of projects, there is much uncertainty. No two projects are ever the same. There are infinite combinations and permutations of outputs to be designed, created, integrated, invented and even discovered.

Because of this, the best that can be done is to create estimates of the time associated with projects. Creating views of a project schedule at gradually decreasing timeframes and increasing level of detail helps deal with the inherent uncertainty.

Skydive from Space

Have you ever seen one of those scenes in a movie or documentary that zooms into our planet Earth starting from space? You know the ones, they begin with a picture of Earth as a blue globe sitting in the black void of space and then the camera zooms in.

Earth gets closer and starts to fill the frame of the picture. Next, we see the upper reaches of the atmosphere and we can begin to make out the shapes of continents. Getting closer, individual countries are visible, then forests, mountains, rivers and cities. Keep going and we can glimpse highways, trees, streets, houses, cars.

Further still, the image takes you in on a small woodland park and zooming in we see a field, plants, flowers and a pond. On and on you go amazed at individual blades of grass, insects that look like giants, drops of water that could fill swimming pools. Switch on the microscope now and you begin to uncover the surface details of tiny bacteria. Inside the bacteria, you are amazed to reveal they are made up of cells, molecules and atoms.

This vision is a good analogy describing how I handle scheduling on agile projects. I begin with a high-level view of the project from start to finish. The initial view serves the purpose of saying to everyone "this is how long our project will run for, details to come later".

We have already created a version of this schedule back in our intent statement. Refer to the section in Chapter 12 that listed the project timeframe. This is our starting point; our photo of planet Earth. We now continue our dive into the project schedule from the outer reaches of Earth's atmosphere and zoom in from there.

The Summary Schedule

I am often amazed at the effect a timeline diagram has on people's perception of what is possible on a project. Without a diagram, imaginations create all sorts of wild and wonderful scenes of outputs that can be created in phenomenally short periods of time. "How hard could it be?" is the catch phrase that sends shivers down the spines of project teams. As you begin to show the actual work in a visual format on a timeline, stakeholders come back to earth and expectations become grounded.

The Agile Summary Schedule is a tool I use as the first step in grounding expectations for a project.

Download a copy of the Agile Summary Schedule from https://miller.productions/agile-toolbox/

A Scrum project is divided into three stages:

1. Initiation and Setup. This includes everything that happens early in the life of a project such as conceiving the idea, forming the project team, determining the plan, setting up the initial working environment, etc.
2. Output Creation. Once everyone on the project agrees on the goals and the associated plans for achieving those goals, the team gets to work. During this stage, outputs are created via a series of iterations.
3. Transition and Closure. Following the last iteration, the final set of outputs are handed over and transitioned into their operational environment. The project is wrapped up, everyone celebrates and all activities reach their conclusion.

This is the first view you create on your summary schedule. A simple timeline, divided into these three stages.

Depending on the complexity of each project, you will see different durations given for each stage. On our conference project, let's make a simple assumption that our project initiation and setup stage will go for three weeks. That will give us enough time to get the necessary people together, form the team and confirm the goals and plans.

We shall also assume that three weeks is sufficient for transition and closure. That provides a week or so to setup the actual conference event on site, a few days to host and run it, then a few more days to celebrate, pack-up, pay any outstanding bills and then close-out project activities.

Identify Iterations

In agile projects, the key element relating to time and schedules is the iteration.

Iterations in agile projects are the individual blocks of time during which the team produces a set of outputs. In Scrum projects, these iterations are referred to as *sprints*.

The output creation stage will involve the team completing several iterations. How many iterations there are depends on the length chosen for each one and the overall time available until transition and project closure.

Between each iteration, there is a break of a few days. This helps give the project team some time to tidy up from the recent round of output creation. It also enables the team to plan what they will focus on for the next iteration.

Iterations also serve the purpose of giving the team regular short-term goals to aim for. For each iteration, there is a set of outputs the team plans to create. At the end of the iteration, key stakeholders gather to review what the team members have completed.

When deciding on the duration for each iteration consider the following:

- Speak to the project team and find out their preference.
- Consider what duration the rest of the organisation and stakeholders are comfortable with. If they are new to agile, working in iterations might be more comfortable with longer timeframes.
- Aim for the *goldilocks* zone. That is, not too short so as to impose too much administration overhead. And not too long such that the project drifts into a more traditional waterfall management style.

- Don't get hung up on bookending iterations into traditional calendar events. For example, don't try to fit iterations to start on Mondays and end on Fridays. Just start and end them whenever they fall naturally.

On most of the projects I run, the duration we end up settling on is three weeks. This provides enough time at the start for the team to get familiar with the work. It also leaves enough time at the end for the team to wrap up their work and get ready for the demonstration. The remaining time, around two-and-a-half weeks is sufficient to gain momentum and complete a decent set of outputs.

I have tried shorter iterations, such as two-weeks long, however this only left a little over one week of solid output creation time. The rest was taken up by ramp-up, wind-down and associated break administration. There is a limit to how much a team can produce with only one week or so. It also doesn't leave much room for dealing with unexpected issues that come up from time-to-time.

As I alluded to above, any longer than three or four weeks sees project teams reverting to those old traditional project management ways. You end up with teams disappearing for extended periods while they produce outputs. It also defeats the purpose of agile since there is less opportunity for stakeholders to provide feedback.

Once you decide on your preferred iteration length, you can update your summary schedule diagram. This time, you will be dividing up the output creation stage and representing the iterations. For each iteration, simply mark it with a time block showing the start and end. Between each iteration, leave a gap for the break.

The Conference Project Schedule

Now that we have determined the times to set aside for each of the three stages, we are ready for the first real version of our summary schedule to go out to the project stakeholders. This information helps visualise what activities are happening and when.

I invite you to treat this next part as an exercise. Have a think to yourself about what the schedule for the conference project would look like. Think about when it would start and how long you would need for the initiation stage.

Next, assume that we allocate three weeks at the end to setting up the conference venue, hosting the event, packing up afterwards, and closing out the project.

Let's also assume that we run iterations of three weeks duration and that there is a break of two days in-between each iteration.

Given this information, how many sprints would you have available? Have a go at drafting up a summary schedule for the conference project.

At this point in the project, you are ready for others to start getting involved on the project themselves. The first round of planning is essentially complete and it is time for others to start producing their own associated project material.

Before they do, however, you are going to establish one extremely critical element. A central place, supported by a single technology, for others to capture and record the information and knowledge they are about to create.

Chapter 16
Capture the Knowledge

Take a moment and flip back to the Agile Manifesto we covered in Chapter 1. Specifically, re-read the part that mentions the value of working outputs over comprehensive documentation.

Once you have finished with that, have a quick look back at the history of agile project documentation we discussed in Chapter 12.

By now, you should be developing a good understanding of the principles behind agile documentation.

On agile projects, we aim for simplicity, succinctness and flexibility in documentation. Conversations between people are the preferred mechanism for communication and information sharing, preferably face-to-face. Documentation should exist mainly to support conversations that help produce outputs.

At this point, things are about to get a whole lot busier. Initial introductions are now complete. People are going to start gathering and producing information relevant to the project. It is going to come thick and fast, in varying formats.

Up to now, you have only created a handful of information products. It would be quite easy for you to store and manage these in your own preferred way.

However, you need a solution that scales to support larger numbers of people working collaboratively together. Now is the time for you to establish systems, processes and tools that others can use to create, store, manage and share project documentation.

Documentation, the Agile Way

In this section, I take you through some of the things to consider in relation to documentation on agile projects. There is a balance to strike between the old-fashioned approach of formally documenting everything in well formatted publications and going completely minimalist with barely anything written down.

Single Source of Truth

Let's begin with the most important aspect first. It is critical that you follow the practice of having a single source of truth. Documentation should only ever be written once and stored in a unique location.

Single source of truth is the practice of storing information in only one place. Never duplicate material.

If a piece of information is used several times, refer to a view of the primary source, don't make copies.

It should be quite obvious why you would want to do this on a collaborative project with multiple people working together. The reasons include:

- Improving efficiency. By only having a single, agreed location to store certain types of information, everyone on the project will know exactly where to go first time.
- Minimising confusion. There is only ever one primary source for each piece of information. The issue of multiple copies is essentially avoided.
- Improved traceability. If anyone ever needs to trace back to the underlying reason or requirement for an output design or project decision, simply refer to the source.

Minimise the Number of Tools

As time goes by, we see more and more technologies being invented to help projects manage their information. Ideas come and go, technologies go in and out of favour. What was recently considered cutting edge, could easily become passé in no time.

The impact of this is that project team members have varying preferences as to which technologies they use to manage their information. If you left it open and free as to how information was managed, you would see a proliferation of tools.

To avoid this, and to minimise the number of tools used for information management, some firm guidelines need to be set.

1. One tool per context. That is, for each specific context or purpose relating to information decide which tool will best suit that need. Once a tool is chosen, use only that tool.

2. Don't overlap. This basically follows from the previous point and aligns with the single source of truth principle. Don't spread information across several tools or locations. Keep documentation contained in one spot so that the reader can find out what they need without jumping around.

Write Just Enough

Remember the key point I made back in Chapter 12? That is, if something must be written, keep it simple. Always keep this in mind whenever creating documentation.

Agile projects value outputs over documentation. Documentation should be there simply to enable and support the creation of outputs. Writing information down in document form serves two purposes on a project:

1. Recording ideas in a permanent location that can be referred to at any point.

2. Ensuring a common understanding among all project stakeholders.

For any idea that is written down, as soon as it provides sufficient knowledge to enable everyone to use it, leave it at that. Don't be tempted to elaborate further. Move on.

Resist the Urge to Continuously Update

Over time, a piece of documentation that was created to serve a purpose goes out of date. In other words, it passes its useful life and is no longer as important as it once was.

On projects where the goal of documentation is to help people produce outputs, obsolescence usually starts occurring around the same time as the associated outputs are nearing completion.

Eventually, documentation gets outdated.

Once a project document gets close to serving its purpose, it can essentially be treated as archive material. That is, consider it a historic record of the project rather than an active document.

Considering this fact, it makes no sense on a project to continuously revisit documentation and keep it up to date. Remember, the goal is to produce outputs, not documents.

If there is a glaring issue in a document, then sure, update it. However, you would provide more value to an agile project by resisting the urge to update documentation all the time.

If the purpose of a document has served and the associated output can be produced correctly regardless of the document's current form, then updating it is only a waste of time and effort.

Share Everything, Where Appropriate

Look back at those previous two points about the purpose of writing information down in document form. The intent is for it to be shared. There is no point restricting it just to a few people on the project, that defeats the purpose.

It is better to share project documentation with all stakeholders by default. That way everyone has an equal opportunity to be informed about any aspect of the project they need to.

For information that can't be shared, i.e. sensitive material, then by all means, restrict it only to those that need to know. However, only do this if there is a genuine reason.

Note that in this section, I have mentioned sharing in the context of project stakeholders and not the general public. If you wish to share you project information publicly, or if there is a mandate to do so, then again feel free. Just be a bit more careful with this one. These days, you never know what a naughty individual could do with snippets of information they have pieced together over time. Depending on what your project is creating, you could inadvertently be putting your stakeholders at risk if you are not careful about information you share with the wider public.

Tool Selection

There are three broad categories of tools in the context of agile project teams. They are:

1. Knowledge management, i.e. writing down and sharing ideas, thoughts, designs, specifications and notes;
2. Task management, i.e. organising, managing and tracking work undertaken to produce and refine outputs;
3. Other tools specific to the industry, profession, project and job at hand.

We won't go into a discussion about the third category, specific tools, given there is essentially an infinite number of directions it could take. That isn't the purpose of this book. However, let us take a brief look at the first two categories.

Knowledge Management Tools

In relation to knowledge management on agile projects, gone are the days of formal document-centric interactions.

Throughout a long phase of history, people have written their ideas down on paper. It was only natural then that project teams did the same. For years, creating a document was the prime choice for capturing ideas. Initially documents were created with pen and parchment, then with typewriters and eventually with computers.

Fast-forward to the 21st century and project teams have broken free of the shackles of documents. Anything that needs to be written these days can be done electronically using software applications that support blocks of text laid out in whatever structure suits the flow and context in which it is used.

The right tools for knowledge management on an agile project are those which enable project team members to:

- quickly write down ideas;
- layout, re-order and shuffle things around;
- refer and link back to other related material;
- leave comments either in-line or in the same space;
- include a variety of formats such as text, images, videos and diagrams;
- create models to represent ideas.

One such type of tool that supports these requirements is a project *wiki*. A wiki is a website that allows users to add, change or delete content using a web browser. The users do so in a collaborative manner and they can structure the content in whichever way suits their needs.

Over the years, several project teams I have worked with have shown a preference toward wikis due to their flexible and unrestrictive nature.

There are other options out there to assist with knowledge management but nothing I have seen beats wikis, yet, in terms of flexibility and ease of use for team members.

Wikis I have used include:

- Confluence (www.atlassian.com/software/confluence).

  This is my current preferred system. It is very easy to use and can be installed locally or cloud-based. It has a range of additional plug-ins that enable functional extensions. It also integrates well with its associated task management tool, JIRA, which I'll mention shortly. Confluence is a proprietary system available commercially for a fee.

- MediaWiki (www.mediawiki.org).

  I learned about MediaWiki back in late 2001 early 2002 shortly after its release. It is an excellent system for capturing information and sharing knowledge. There is a famous site on the internet you are most likely familiar with that is based on the MediaWiki platform, Wikipedia. Enough said. MediaWiki is free, open-source software.

- Foswiki (foswiki.org).

  Foswiki is based on one of the earlier wiki platforms, TWiki. It has similar capabilities as MediaWiki and a clean design.

There are many other wikis out there, too many to cover in this book. A good resource exists, WikiMatrix (www.wikimatrix.org) that provides a comparison between the main platforms should you have a need to choose.

Lastly, before concluding this section on knowledge management tools, there is something you should be careful of.

When considering knowledge management tools, only use one for the project.

Don't be tempted to mix several knowledge management tools together. Doing so will only leave you with information all over the place. It will eventually be very hard to find things. Your information will turn into a chaotic mess which nobody can use.

Task Management Tools

The next category of tool we consider for agile project teams is one to organise, manage and track the work of producing outputs.

Within this category, agile teams benefit specifically from having a tool that supports the *plan-output-review* cycle. That is, a tool that enables team members to:

- create items that represent individual blocks of work;
- see the entire set of possible work items for the project, i.e. the backlog;
- raise or lower the order of each work item in terms of priority;

- create subsets of work items which the team will estimate and work on during an iteration;
- reflect the status of work items as they progress from an idea, to being worked-on, reviewed and then finally completed;
- assign work items to individual team members;
- record conversations among the team about the work.

The types of tools that support these requirements are known as project task tracking or project management tools.

You will absolutely want a task tracking tool on your agile project. Without it, the project team will be flying blind or just winging it so-to-speak. In other words, work will be undertaken organically without any visibility or coordination. Agile projects run best when there is order and structure to work undertaken.

Ideally, the tool will have built-in support for the flavour of agile you are working with. For example, in our hypothetical conference project, we are working with Scrum. A tool that had built-in support for Scrum activities would be helpful.

Task tracking, a.k.a. project management, tools I have used include:

- JIRA (www.atlassian.com/software/jira).

  This is my current preferred tool. It focuses around items of work, known as issues and their lifecycle through a workflow. It is extremely customisable, so you may label your issues, define your own custom workflows and tailor the attributes of issues to suit your project. JIRA has out of the box support for Scrum, particularly in terms of backlogs and sprints. It integrates very well with Confluence.

- Basecamp (basecamp.com).

  Basecamp is one of the earlier online project management tools that became popular. Its support for remote team member collaboration on project tasks was quite unique for a while. Basecamp is known for being simple to set up and start using. It has a strong following with freelancers and small business.

- Microsoft Project (office.microsoft.com/project).

  One of the more commonly known project management tools, Microsoft Project, is excellent for traditional predictive planning project management. It has many powerful features that could enable one to have quite a fine-grained view on all aspects of a project. This focus on predictive planning means Microsoft Project it is not so useful for agile project management at present.

- Oracle Primavera (www.oracle.com/primavera).

  Primavera is similar to Microsoft Project in its focus toward traditional predictive planning. It is aimed more toward the enterprise / large corporation market.

- Trello (trello.com).

  One of the more basic task management tools, Trello lets you create and view lists of tasks organised into columns. Being web-based it can support remote team members. Aimed at smaller teams and simple projects, it does this very well.

This list merely scratches the surface, there are way more tools out there. Take a look at the Wikipedia page titled *Comparison of project management software*[xv] to see a summary of many of them.

Once again, when deciding on a task tracking tool for your project, pick one tool and stick with it. Don't use more than one tool for this purpose. Tracking work in more than one location is a recipe for disaster, conflict and all-round confusion for project teams.

Lastly, you certainly don't want to make it optional to use the project's designated task tracking tool. Agile projects need focus and momentum to keep pumping through the outputs. Project teams should only be working on items that are in the project backlog and they should do so in priority order. Anything outside of that is questionable in terms of supporting the project's goals.

Now that we're successfully bringing together the information and knowledge for the project, it's about time we started making some progress on the work of creating actual outputs.

Chapter 17
Organise the Work

At this point in the project, we are still in the initiation and setup stage. We have a budget, a schedule, and an initial set of systems, processes and tools.

Our goal now is to get the team to a point where they can take responsibility of the work and start producing outputs.

In this chapter, I take you through the steps of identifying the outputs that are to be included in scope. Following that, I demonstrate how to establish the Scrum processes associated with sorting the work. It's time to start focusing on the backlog.

What Are We Building?

The questions in focus are: What outputs should the team produce? What are the stakeholders looking for? What does the project owner want? What do the customers need?

The first place to start answering these questions is to look back at the target outcomes and benefits the project is aiming for. This is what it is all about. These goals are why the project exists.

Whenever there are questions such as "what are we building?", thoughts should include target outcomes and benefits. Having these goals makes answering such questions easier. If an output contributes to target outcomes and benefits, it is possibly in scope.

Start with a Work Breakdown Structure

A Work Breakdown Structure, or WBS, is one of the best tools for getting people on the same page about what outputs a project is going to produce.

Early in the project, everyone has their own individual ideas about the upcoming work ahead. This is great for innovation and inspiration, but at the same time potentially leads to fragmentation.

Some people come up with thoughts for outputs early. Their imagination leads them to designs and solutions. Other people think more in terms of tools and environments. That is, setting things up so that work can commence. Further still, folks may think in terms of phases or activities. Going even broader, some people think about the various skill specialties that will be needed to produce outputs. Others think more in terms of the personal relationships.

Without a unifying approach, you can see how these different views could result in people going off in various directions.

A WBS does the job of bringing all these ideas together and representing them in a way that focuses everyone cohesively on what will be delivered.

A Work Breakdown Structure, WBS, is a way of visualising the main deliverables of a project.

Starting with a few top-level deliverables, each one is divided down into its component parts. This is repeated successively until logical and manageable work packages can be identified.

The most common form of a WBS is a diagram that uses boxes joined by lines, displayed in a hierarchical manner. The top box represents the project. The next level down, connected by lines from the top, is a small set of elements that represent the main deliverables for the project. Beneath each of these elements are further items that represent the components of the main deliverables, and so on.

How Far Down Does a WBS Go?

Deliverables in a WBS continue to be decomposed until they reach logical terminal points. In theory, a fully complete WBS keeps going down until it reaches a set of elements that could each be picked up and worked on easily and independently by an appropriate person or sub-group within the project team.

Keep in mind that in an agile project we have a preferred approach to documentation: write just enough. Just enough in this context relates to the creation of user stories.

In an agile project, the lowest level elements in a WBS should be just enough that a project team could visualise all outputs together and then begin to create the associated user stories.

Remember, agile documentation should aim to facilitate the creation of outputs. A good WBS in an agile project lets the stakeholders have a shared understanding of the main outputs and the component parts. From there, the team can use the knowledge to get on with the job of defining the associated user stories and working with them to create outputs.

Who Creates the WBS

To create a WBS, you ideally want the following people involved:

- One person to facilitate its creation and keep it updated, usually the project manager.
- All interested stakeholders to provide ideas and thoughts.
- Project team members to verify the agreed version.

The key is the involvement of the project team since they will be designing and producing the outputs listed on the WBS. Feasibility, ownership and professional opinion are the main things to look for.

WBS Tools

A WBS could take various forms, including:

- a simple list with indentations for levels;
- boxes with interconnecting lines;
- spreadsheet with columns and rows;
- organisation chart;
- mind-map.

If you consider that a WBS is simply a representation of a hierarchy displayed in tree form, then you can produce a WBS using any software tool that has that functionality.

Ideally, the tool will be purpose built and have support for WBS-specific functions. That way, you can create a WBS in rapid time with minimal effort and keep it updated easily.

Using software that was built for other purposes, but that could be stretched to provide WBS-like capabilities, is reasonable. However, you might find that trying to be tricky with that other software to get WBS-like support could keep you less productive, and possibly frustrated, over the long run.

WBS tools I have used include:

- WBS Schedule Pro

  (www.criticaltools.com/WBSChartPro.html)

  This is my current preferred tool for creating WBS diagrams. It is custom-built specifically for creating and maintaining work breakdown structures. Features include simple drag-and-drop, connecting, aligning, numbering, layering (i.e. hierarchy nesting), rolling-up elements (i.e. show/hide), data fields for elements, summary views, focusing on individual elements. It also has extra functionality to use for scheduling purposes and integrates with Microsoft Project, however I don't use those last two features with agile projects.

- Microsoft Project (office.microsoft.com/project)

  This software does a good job at producing a WBS in spreadsheet form. It has built-in support for labelling, numbering and layering. Modifying and updating WBS elements is flexible and easy. One issue though is that it doesn't display a WBS in a diagram form (i.e. boxes and arrows). For that, you need to export out to another software package.

- Microsoft Visio (office.microsoft.com/visio)

  Visio does a good job at creating a WBS in diagram form. Given Visio is a diagramming tool by nature, it does a lot of things well in the context of producing a WBS, i.e. alignment, linking elements, resizing, etc. There are also WBS-specific plug-ins and templates available that help enhance Visio to provide additional WBS capabilities.

The Backlog

Now that there is a common understanding among stakeholders about the main outputs of the project, it is time to start getting down into the details. We are ready to start creating agile user stories and building up our project backlog.

Preferably, if the budget of your project allows it, you would begin working closely with someone at this point that has strong requirements analysis skills. This person is going to help identify, record and manage the details of all the agile user stories for the project.

Trust me, you are going to want a dedicated individual in this role.

Agile projects can have anywhere from a few hundred to a few thousand user stories created in their backlog. While the project may not end up producing all those stories, they still end up in the backlog simply due to the continuous flow of ideas.

To the Stakeholders We Go!

Anytime you hear the words user stories, they should be a trigger to check that the project stakeholders are involved in some way.

Including the thoughts and opinions of the people using those outputs to create the associated user stories is the key to success. There is no point letting the team assume they know what to build. They could get things right. However, they could just as easily get things wrong.

It is obviously too unwieldly to try and interact with all stakeholders all the time in relation to the creation of user stories. For that reason, we usually go with a representative person or group. That is, one or a few people that are nominated to represent the views of certain stakeholders.

Back in Chapter 4 we described the project board as a way of bringing together stakeholder representatives.

The First Users Stories

A simple way to get started with producing user stories is to:

- Refer to the WBS to identify high-level components or categories.
- Ask the stakeholders what they would like to see first. That is, ask which outputs would best help the stakeholders begin to understand and get a feel for what they need from the project

The response you get will vary and there will likely be a sense of uncertainty early on, but that is OK. Agile project management is designed to help bring people along that journey from uncertainty to value and hopefully even delight.

Another area I like to focus on is any set of outputs that are complicated or challenging for the project team. If an output is going to be hard to produce, it is better for the team to get a start on it early rather than later.

That way, the team members have plenty of time to work their way through the issues and come up with suitable solutions.

Anatomy of a User Story

Put simply, a user story describes a function or capability of an output that helps a customer achieve a goal.

A user story is written using the following sentence structure:

As a *customer of some type*, I want a *function or capability of an output*, so I can *achieve a goal*.

For example, in our conference project, one user story could be something like: as an attendee, I want food to be catered, so I can concentrate on the event.

In addition to the description, there will be a set of details and criteria that define what makes the user story fit for purposes.

That is, what would make it acceptable to the customer?

These could include anything relevant to the story itself. They could be certain features that should exist, or certain constraints that need to be considered, or even things that should be avoided.

Along with a user story there should be a set of details or criteria that describe how to make the output acceptable.

For example, following our conference project story regarding food catering, the acceptance criteria could include: must support special dietary requirements, should have a range of light meal types for lunch, tea and coffee options to be provided, healthy snacks available between meals, avoid foods with high sugar content.

The Ideal Size for a User Story

The sweet spot for user stories is to break each one down to a point at which it could be completed in reasonable time. On most of the projects I run, that timeframe is anywhere from one or two days to around one week or so.

Notice I am being quite fluffy in my description there. That is deliberate. Some of the time it is possible to define a user story that can be completed by one or two people in just a day or two. Other times, there will be complicated items that take longer.

For complicated items, it often doesn't make sense to break them down into smaller timeframes just for the sake of it.

The reason we use time as a defining factor for the size of user stories is that Scrum is primarily managed around schedules. There is a fixed schedule for the project.

Within that schedule there are a set of sprints.

Within each sprint you are aiming for a suitable number of stories to be completed so that you can get useful feedback from stakeholders.

If the stories are too small, they won't achieve a logical goal for a customer. Too long and the project team won't deliver much in the sprint.

Write User Stories in the Task Management Tool

The next step as you begin to create user stories is to get them into the nominated project task management tool.

The single source of truth for user stories is preferably the task management tool, as outlined in Chapter 16. User stories are the building blocks of a Scrum project, from which everything will be built.

Work will flow on the project around user stories. Thus, it makes sense to write them where they will be used.

WBS & User Stories for the Conference Project

Conference Project WBS

Let's see a WBS in practice by creating one for our example conference project.

The top-level elements in a project WBS are very general representations of the main deliverables. The idea is to start with simple broad concepts that help people focus on each category of output.

Each top-level element is then divided down, or decomposed, further into is component parts.

In our conference project, a hypothetical set of first and second level elements in the WBS, i.e. the main outputs, might include:

- Venue
    - Rooms;
    - Equipment;
    - Catering;
    - Security;
    - First Aid;
    - Facilities;
    - Decoration.
- Logistics
    - Dates;
    - Bookings;
    - Schedules;
    - Runsheets;
    - Support Staff;
    - Procedures;
    - Materials;
    - Co-ordination.
- Marketing
    - Website;
    - Social;
    - Press Releases;
    - Advertising;
    - Subscribers;
    - Direct Communications.
- Exhibitors
    - Application;
    - Registration;
    - Policies;
    - Floor Plan.

- Speakers
    - Candidates;
    - Invitations;
    - Submissions;
    - Biographies;
    - Synopses;
    - Timetable.
- Sponsors
    - Sales;
    - Tiers / Levels;
    - Promotion.
- Attendee Experience
    - Ticketing;
    - Communication;
    - Program;
    - Event Packs;
    - Networking;
    - Entertainment;
    - Prizes;
    - Feedback.

Conference Project User Stories

Based on the high-level elements in our WBS we have a good starting point to get ideas for the first user stories. Next step would be to speak to the stakeholders and find out what would provide them most value to help them visualise what they need from the project.

Let's assume you do this and the consensus is that knowing the date for the conference is one of the most important items. Based on this we can produce the first user story.

User Story 1: Conference Date

As an organiser, I want to know the date of the conference, so I can organise other aspects of the project that depend on it.

Acceptance criteria:

- Sufficient lead time.
- Doesn't clash with other major events.
- Is reasonably achievable.

User Story 2: City

As a participant, I want to know the city for the conference, so I can make plans that depend on it.

Acceptance criteria:

- Close to a major international travel hub.
- Good local transport options.
- Suitable accommodation supply.
- Enticing tourism factor.
- Low security risk.

User Story 3: Venue

As an organiser, I want a venue, so I can run the conference.

Acceptance criteria:

- Sufficient capacity.
- Aligns with budget estimates.
- Suitable facilities.

User Story 4: Speaker Candidates

As an organiser, I want a list of possible speaker candidates, so I can start approaching the ones most suitable.

Acceptance criteria:

- Align with the theme of the conference.
- Sufficient experience.
- Notable community profile / personality.
- Would provide value to attendees.

And So On...

As you can see we have merely touched the tip of the iceberg in terms of user stories for the conference project. These are just a few examples to get you thinking. Let your imagination run to think about all the other outputs that would need to be produced.

From this starting point, the process of defining further user stories continues. It will be a gradual iterative process. Each time a new user story is created, the backlog for the conference project grows.

Congratulations, you should now have an initial WBS and backlog of user stories that define the first set of work for the project. We're almost ready to hand things over to the team to start working. Before we do, however, let's go through that all important agile activity, prioritisation.

Chapter 18
Prioritise

Prioritisation is the art of putting the user stories into order of more important to less important. The more important items are worked on as soon as the team is ready. Less important items wait until others ahead of them are completed.

Prioritisation provides significant value in an agile project. Stakeholders are given regular opportunities to adjust the sequence in which outputs are produced. This provides a good fit to a naturally evolving understanding of what people need from the project.

Projects are so much more comfortable and predictable now that agile prioritisation has arrived, compared to what it used to be like.

Trust Us, We're the Professionals?

I can recall what it was like before agile project management turned up on the scene. Conversations were a bit tense at times. Project teams would spend significant effort up front on the project trying to document requirements of all aspects of the project in detail.

There would be numerous workshops and meetings with stakeholders who, naturally, would have different views and ideas of what they would need.

It was difficult for anyone to get a mental model of the final outputs. Unless you have seen something before, it is hard to visualise it, and even harder to imagine using it.

Eventually, the project team would produce the *agreed* set of requirements. These documents resembled the epic novel war-and-peace in terms of length. The project clients were urged to *sign-off* the requirements to confirm they were an accurate representation.

After all that was said and done, which mind you could take several months, the team would disappear and start the job of building the outputs. There would be very minimal interaction with stakeholders from this point onwards.

At the same time, there was very little opportunity for stakeholders to provide feedback or update their priorities. Such activity was discouraged and considered a *change request* that had much stigma associated with it.

Finally, the project team would rejoice, with much fanfare, on completion of their mighty effort. Metaphorical fireworks would be let off when the outputs were delivered on-spec, i.e. achieving all the documented requirements. Further congratulations and back-patting would be seen if the project was completed on-schedule and within budget. Success all around.

There was only one problem. Quite a big one, actually. Slightly inconvenient. Even though these projects were a success on paper, they often failed in other ways. There was often no real achievement of target outcomes. And because of this, the benefits that everyone was hoping for failed to materialise.

There is a Better Way

The folks that brought agile prioritisation into the mainstream have so much to be thanked for. They have made it possible for a more logical and natural way of producing outputs to be undertaken.

Creating novel products and services in projects is hard. It requires bright minds, innovative thinking and intricate problem-solving ability. For the folks on the receiving end of the outputs, it is a better fit for human cognitive skills to gradually build-up an understanding of each new item that is produced. While people can conceptualise things, our minds are better suited to work with our sensory systems.

Prioritisation enables the complexity of dealing with a large list of user stories to be simplified down to focus on the things that matter. Details of other lower priority items can be dealt with later.

Stakeholders can regularly check-in to physically see and use outputs as soon as they are working. They can gradually improve their understanding each time. The most important result of this is stakeholders being able to provide feedback and update their needs based on their improved understanding.

Once the first set of outputs is delivered, stakeholders really get a sense of the end-game. They get excited, involved and energetic. Here is a project that is listening to them. Letting them adjust the priorities. Can you believe it, the stakeholders are the ones in charge!

The time has come to trust them.

Enter the Product Owner

The main person we work with to prioritise the backlog in Scrum projects is the product owner. Refer to Chapter 13 where we introduced this role. The product owner is the person that represents the views of the key stakeholders in respect of the outputs being produced. The product owner is responsible for helping define the priority order of the outputs in the project backlog.

Consider Their Point of View

Prioritisation is serious business and it can be quite challenging at times for the product owner. You see, they essentially end up in a role that by its very nature can carry a lot of weight and expectations.

The product owner is speaking on behalf of several people on the project. The success of the project, in some respects, sits partially on their shoulders.

I have observed several emotions in product owners on Scrum projects. At times you see comfort, ease and clarity. While other times, there is confusion, anxiety and occasionally stress. They worry about getting the priorities right.

Help Them Get Rolling

The best thing you can do regarding prioritisation is to put the product owner at ease. You will be undertaking prioritisation together regularly on the project and they will eventually get used to it. However, in the first few times, there could be genuine concern.

Do your product owner a favour by helping them any way possible with the earlier prioritisation sessions. This could include:

- meeting with them to personally talk through the initial set of candidate stories that have been generated;
- developing visual-aids such as diagrams, mock-ups or models so they don't just have to rely on their imagination;
- reassuring them that they don't need to get things perfect since the Scrum process is intentionally flexible;
- describing what activities and actions will happen following each prioritisation session, paying special attention to the agile *plan-output-review* cycle.

When to Prioritise

There are two ways to answer the question of when to prioritise.

On one hand, you should run prioritisation sessions with your product owner as much as they feel necessary. The goal is to ensure they feel the order of outputs in the backlog matches the views and opinions of the people they represent.

On the other hand, you want to run prioritisation sessions at regular and consistent intervals. That way it becomes an expected and planned activity. Being methodical about prioritisation helps safeguard from the rare cases where a product owner wants to change priorities either too much or too little. That is, they keep changing their mind, or they get distracted by other responsibilities and disengage from the process.

On Scrum projects, I encourage one or two *light* prioritisation sessions early in the project. I then aim for a regular cycle of one prioritisation session coinciding with each iteration, i.e. in line with each sprint.

Following each sprint there will be a demonstration of the new outputs produced. This will be an opportunity for stakeholders to review the delivered work and provide feedback to the product owner. This new information and understanding often leads to updated preferences about the order of outputs remaining in the backlog.

How to Prioritise

Focus on Target Outcomes & Benefits

Always remember that a project exists to achieve certain target outcomes and ultimately deliver benefits.

Everyone involved in prioritisation should keep this in mind and focus on items which have the highest contribution to those goals.

Encourage Slices Through the WBS

Looking at the WBS, you have a set of outputs, each with a deep set of component parts. The entire set of components and outputs makes up the full scope of the project.

Now, there are several ways to approach the production of these components. The team could work on them randomly or in whatever combination that prioritisation preferences come up with. They could focus on delivering all components on a specific output before moving onto the next output, and so on.

Instead, I like to encourage slices through the WBS. What this means is that the team works on a cross-section of components from every output category.

Taking a *slice* approach, stakeholders begin to see representative parts unfolding from the entire set of outputs. This speeds up the feedback loop. All outputs of the project become tangible sooner.

Take a Broad Look, Then Drill Down

Several prioritisation techniques have been developed over the years. These include:

- Ranking;
- MoSCoW (i.e. must, should, could, won't);
- Bubble Sort;
- Numerical Assignment;
- Analytic Hierarchy Process (AHP);

I often encourage stakeholders to start their prioritisation with a quick broad look over the entire backlog and then to pay closer attention to the items toward the top of the list.

Using the MoSCoW technique is one way of achieving the initial pass over the backlog. Stakeholders can rapidly scan over each item and apply one of the following categories:

- Must: critical for the project to be a success.
- Should: important for success but not necessary.
- Could: desirable but again, not necessary.
- Won't: provides the least value.

Following this, attention can shift to the items rated *must* to get them in a more detailed order. If desired, the prioritisation can move onto the items rated *should* and *could*.

Hang Around at the Top

There are diminishing returns to the prioritisation effort the further down the priority list you go.

The further down the list, the longer it will be before the project team has capacity to work on those items.

There are items in the list that won't be in focus for the team until toward the end of the project, if at all. Hence, there is little benefit from spending too much time analysing those ones.

Instead, a better approach is to focus attention on items up toward the top of the priority list. Get through the higher order items first. Then, once they are complete, re-visit the next block of items in the list in more detail, and so on.

Look for Complex and Challenging Items

Back in the previous chapter, I mentioned the idea of focusing on complex and challenging items first.

Working on challenging items early provides a more accurate estimate of what the team might be able to achieve for the remainder of the project. It is better for everyone to get a realistic sense early about how many items the team can compete in each timeframe.

If you leave the challenging items that take longer until later in the project, your estimates of overall throughput tend to blow out and the team under-delivers on expectations.

Dependencies Matter

On most projects, there will be certain components that must be commenced or possibly even completed before other components can be progressed.

For example, before you pour a concrete slab, you need to put the formwork in place.

Identifying dependencies will impact the order of work and thus plays a part in prioritisation. Some items will, by necessity, have a higher priority than others.

Similarly, it may not be possible for some items to have the priority that the stakeholders desire due to dependency constraints.

Consider the Burndown Rate

At the end of each sprint you get an update on the throughput capacity of the project team. That is, you can calculate the average number of user stories the team is able to produce each iteration. This is known as the *sprint burndown rate.*

The burndown rate can then be used to produce a forward estimate of how many user stories the team is likely going to complete in the remaining sprint iterations left on the project.

Thus, you can get a rough idea overall of what items in the backlog could possibly be completed and items that are unlikely to make it due to capacity constraints.

Visualising what won't be included in a project helps guide the thinking of people involved in prioritisation. It ushers thoughts away from an optimistic mode, i.e. wishful ideas that the project team can deliver what everyone *thinks* they can.

Instead, the folks involved in prioritisation are grounded more towards a realistic mode, i.e. objective measurements based on the observed fact of the actual burndown rate.

Gather Different Opinions

To wrap up this chapter, let's consider one of the most important aspects of prioritisation.

Priorities should represent the opinions of a broad cross-section of stakeholders.

It is imperative that priorities do not just reflect the personal opinions or interpretations of a small number of people. Even the most well-intentioned product owner can inadvertently sway priorities due to their own bias.

The final outputs of the project will eventually be used by many people. It is vital that the actual folks that will be using the project's outputs get a chance to share their opinions. Success depends on it.

Prioritisation for the Conference Project

As an exercise, think about how you would go about prioritising the backlog for the conference project. Who would you involve? What do you think the initial priority order would be?

There is no right or wrong answer to this, so have a go at it. Grab a piece of paper and a pencil, or open a notebook on your preferred computing device if that works for you.

In the next chapter, we look at ways the project team uses the priorities to guide their creation of outputs. This is where the rubber meets the road on Scrum projects. It is time to start the sprints.

Chapter 19
On Your Marks, Get Set, Sprint!

Sprints: where the magic happens on Scrum projects. The sprint is the core structure within which all other activities either lead to or come from.

The sprint is a period of time during which the project team aims to deliver a set of outputs in priority order.

In this chapter, we go through what is involved in running a sprint. We cover planning and estimation, daily stand-ups, task tracking and retrospectives.

Sprint Planning

The first step in every sprint is the planning process. Putting in sufficient time and effort up front makes everything else in the sprint go much smoother.

Our Targets

- A prioritised list of stories to be targeted in the sprint and their associated estimates.
- Enough information for the scrum master to enter the planned activities into the task tracking tool.
- Enough information to enable the team to get started on producing deliverables for the sprint.

The General Approach

For each sprint, we proceed through the following series of activities:

1. Commence preparation.
2. Review the backlog.
3. Commence analysis.
4. Review analysis progress.
5. Pre-planning.
6. Sprint planning.
7. Record the outcome.
8. Kick-off the sprint.

Each activity is described in more detail below.

Commence Preparation

Prepare for planning by booking in meetings and allowing sufficient lead-in time to carry out any necessary analysis (e.g. at least 2-3 weeks lead).

During a Scrum project, I normally run sprints that last for three weeks. To prepare, I aim for the team to have enough analysis completed during a three-week sprint in time for the following sprint.

Ideally, the team would get an even larger head start than this, so they are well prepared and the backlog is well described.

This buffer is achieved occasionally, but on average most teams run analysis three weeks ahead of development.

Review the Backlog

This is where the next deliverables are identified for inclusion in the upcoming sprint. The previous work of prioritisation makes this step quite easy.

Essentially, the team just picks out the number of highest priority items from the backlog that matches their average burndown rate.

Commence Analysis

We now have an idea of which deliverables are going to be targeted in the next sprint. From this, members of the project team commence carrying out the analysis necessary to support production of the deliverables.

Remember how I said earlier that you would ideally have a person in the project dedicated to doing this? Here is where their value lies. The goal is to flesh out the details in each user story to ensure there is a thorough description, acceptance criteria and any other associated material (i.e. models, diagrams and mock-ups). It takes a fair bit of effort and isn't something you can treat as a secondary activity.

Review Analysis Progress

Approximately half way through the current sprint, a small workshop is held to review analysis progress.

Invited to the workshop are the people undertaking the analysis, any key members of the project team and the scrum master.

Together as a group, they briefly review the current set of analysis material that has been produced to ensure it is on track for delivery. The goal is sufficient content in time to be of value to the project team that will be using it. Any items that need actioning are noted.

Pre-Planning

This is an optional meeting for the team to discuss the user stories scheduled for the next sprint and start thinking about them. It is more likely to be used when there is a high degree of uncertainty or complexity relating to the stories.

- Each story from the sprint backlog is presented by the analyst and discussed by the team.
- Together as a group, clarify any points as necessary.
- The analyst also notes down any items requiring further investigation or updating.
- This is also a good time to identify any dependencies which may affect the order in which stories are worked on.

Sprint Planning

The aim of the planning meeting is to identify the set of user stories that will be targeted in the upcoming sprint. While this is a meeting with a high degree of technical detail, i.e. team members discussing how they will build things, it is good for the product owner to attend and take an active role.

- To prepare for the sprint planning session, the nominated scrum master determines each team member's availability for the time period and calculates the total amount of effort hours available in the sprint. They also create a list of the stories to be considered in priority order.
- The planning session then cycles through each user story:
    - The team discusses the story to ensure they have a full understanding of what is involved.
    - Each team member with enough knowledge of what is involved provides an estimate for how long it will take to complete the story.
    - If there are any major differences between estimates for a group, these should be discussed to identify why the estimates vary and come to a consensus.
    - Conclude with an average of all estimates.
- Each time a story has been estimated, it reduces the amount of effort hours left in the sprint.
- Once the total available effort hours for the sprint are estimated to be used up, planning is completed.
- The team and product owner settle on an agreed estimate for which user stories will be targeted during the sprint.

Record the Outcome

The scrum master for the sprint transcribes the information from the planning session into the project's task tracking and knowledge management systems. Essentially, the stories that are to be included in the sprint should be updated to reflect this.

Items that didn't make it into the sprint, are put back onto the backlog.

Lastly, the team members break down each story in scope into sub-tasks to be carried out individually. Each story is decomposed into tasks to reflect how the output will be produced.

Kick-off the Sprint

The final step is for the scrum master to start the sprint. The status in the project's task tracking system is updated, calendars are marked, and the iteration clock starts counting down.

Daily Stand-Ups

Also known as scrum meetings, daily stand-ups are short meetings run by the scrum master to keep team members up to date with progress during a sprint. They are usually held in the morning.

Daily stand-ups are a way for everyone in the team to collaborate and learn together in the sprint. They also help highlight any blockers that are getting in the way of people completing their user stories.

In summary, the daily stand-ups run as follows:

- They are hosted by the scrum master.
- Each participant answers three questions:
    1. What did I do yesterday?
    2. What will I do today?
    3. Are there any blockers?
- Each person keeps their response as short as possible.
- Any issues that result in further discussion are noted and set aside for a separate conversation after the stand-up.
- The total time is limited to 15 minutes or less.

If the project team is co-located, then it is advisable to hold each session in a location where the team can physically stand-up and be free to talk without interrupting others. The act of standing encourages people to remain brief when it is their turn to talk.

If the project team includes people that are in physically separate locations, then it is recommended that everyone dial-in using some form of communication technology that includes voice, video and screen sharing. That way, everyone is on an equal platform and nobody feels *on the outside* of the group.

Task Tracking

An important activity that sometimes gets overlooked in conversations about Scrum is task tracking. As discussed previously, the team enters user stories into the project task tracking tool, so they can be thorough and methodical in how they go about their daily work.

Remember our single source of truth strategy for documentation? Here it is once again.

At the start of a sprint, the user stories in scope should be separated out so the team can easily see what they will be working on. As the sprint progresses, each team member should keep the task tracking tool updated so it accurately reflects the state of the sprint at any given time.

This includes marking whatever they are working on as being *in-progress*, placing items *on-hold* if they are waiting for some reason, and progressing things through to *done* when applicable.

As work proceeds, it helps to record notes, ideas, questions and comments against the applicable records in the task tracking tool. This helps anyone interested to become familiar with a particular item in rapid time, in the one place.

Retrospectives

A retrospective is a private catch-up between team members and held at the end of each sprint.

Retrospectives are a way for the team to self-assess their own activities in relation to the tools, techniques and processes they are using to create outputs during a sprint.

The goal of each retrospective is to produce a set of improvements that can be made to the way they undertake their work.

Two main questions are asked in the sprint retrospective:

1. What went well during the sprint?
2. What could be improved in the next sprint?

It is important to discuss both questions, so participants don't find retrospectives a negative experience.

Team members should each take a turn sharing their thoughts with the others.

As mentioned above, limit these sessions to invited team members only. That way, they can feel comfortable and at-ease with sharing their concerns together. The idea is to foster a spirit of positive continuous improvement rather than seeing it as a critical performance review.

Conference Project Sprint Number One

Exercise time. Refer to the activities we take for each sprint as described in this chapter. Next, visualise how they would be applied to the conference project example.

Sketch down what you think you would do for each activity involved in sprint planning? When would you hold the daily stand-ups? Who would be your scrum master and what type of task-tracking system would they use? How would you run your retrospectives and what sort of questions would you ask?

Once again, there is no right or wrong answer to this. Just be sure to write it down for the experience of applying what you are learning.

In a nutshell, that is all there is to running a sprint in a Scrum project. It is quite simple really and doesn't need to get much more prescriptive than what I have outlined here. The Scrum approach is quite logical and simply provides a layer of structure, formality and regularity to the work undertaken by the team.

Now that the work of producing outputs is underway, it is time to consider sharing progress with our stakeholders. In the next chapter, we open a window on the project and consider collaborative activities such as demonstrations, reviews and feedback.

Chapter 20
Share the Progress

A major theme in *Don't Spook the Herd!* is the importance of people. This idea has essentially been up on a giant 18m x 6m billboard with a high-powered spotlight shining on it throughout the entire book. In Chapter 4, I stated that projects are carried out by people for people.

Given we're comfortable with this idea, why not revisit it here for good measure? In case you didn't notice, I'm a big fan of this idea.

People are the key to successful projects.

Why come back to this point with such energy in this chapter? The reason is simple: now that the project is running and the team are producing outputs, it is more important to be engaged and working closely with stakeholders than at any other time in the project.

In this chapter, we look at the various ways we engage with stakeholders on Scrum projects to share the team's progress. We begin with a progress sharing strategy and then discuss various tactics to implement that strategy. Along the way I'll also cover how to deal with various issues you might face when sharing progress with large groups of stakeholders.

Progress Sharing Strategy

The strategy to follow in agile projects comes back to the *items on the left* in the Agile Manifesto, namely:

- individuals and interactions;
- working [outputs];
- customer collaboration;
- responding to change.

Now that the project is producing working outputs, we are ready to start delivering on the other elements. We implement this strategy through the following tactics:

- reference groups;
- sharing project content;
- sprint demonstrations;
- output reviews;
- managing feedback.

Reference Groups

A reference group is a collection of people that represent a certain category of stakeholders for the project.

We have already formed and discussed one special type of reference group in the book so far. That is, the project board. This group is special in that it not only exists for reference purposes, but it also takes on the responsibility of governing the project.

There are a couple of others worth considering also. The other two reference groups I normally recommend working with are:

- user / customer reference group;
- subject matter experts.

User / Customer Reference Group

This is a group of people that represent the end users or customers of the project's outputs.

In other words, the actual people that will be utilising the final outputs once the project is complete.

Forming a group like this and sharing progress with them provides so many positives to the project. It is through their use of project outputs that target outcomes will be generated. The eventual benefits of the project are closely linked to how well the project outputs fit the needs of this group.

From a change management aspect, having a user / customer reference group is also constructive. For projects that have a high need for change management, bringing people along for the ride and having them part of the project is a technique that is often suggested.

Tips to forming a user reference group and working with them include the following:

- Identify specific people that will eventually be utilising the outputs of the project.
- Inform them of the strategy of involving them regularly throughout the project to share progress and get their feedback.
- Invite people to voluntarily join the group. Let them know their commitment is limited to only a few hours every two to three weeks.
- Emphasise the value to them for participating is that they will end up getting outputs that are more closely aligned to their personal needs. The more they participate, the more likely the outputs will be a good fit for them.

Subject Matter Experts

A subject matter expert, or SME, is somebody considered highly knowledgeable in some aspect relating to the project. SMEs usually hold several years of experience and training in a certain field.

A project team can discover amazing gems of wisdom by sharing their progress with subject matter experts and inviting feedback.

I try to seek out and arrange for SMEs to be involved in my projects whenever the team is working on particularly challenging items.

The challenging part about working with SMEs is that they are usually quite busy with their own work outside your project. Occasionally, you might be lucky enough to have them as dedicated project team members. But it is more likely they will be invited to collaborate for short stints.

Some ideas for working with subject matter experts are as follows:

- Inform SMEs of the problem faced by the team and invite them to share their expertise to help with solutions.
- Provide context about the problem in terms of the overall timeframe and scope of the project. It could be a short specific hurdle to overcome, or it may be a long running aspect that impacts many other areas in the project.
- Give SMEs a general idea of what you are looking for from them. That is, what you would like them to do. Are you simply seeking their opinion, advice and guidance? Or do you need them to actively help design and produce solutions?
- Arrange for the applicable members of the team and SMEs to collaborate.
- Follow-up on the results and outcomes.

Sharing Project Content

We mentioned the art of sharing project information and documentation already in Chapter 16. Here we reinforce that idea of sharing everything, where appropriate.

The tactic to put in play is publishing regular updates out to interested stakeholders. Think of it as a combination of marketing, promotion, updates and status reporting for your project.

In addition to making available the project information discussed in Chapter 16, the goal here is to take a pro-active approach to keeping stakeholders informed. You are aiming to provide them with interesting content that keeps them excited about the project.

When deciding what to share and the language or tone to use, consider the specific needs of your own project. However, I would thoroughly recommend treating this aspect of agile project communication with a degree of fun and entertainment.

Think of things from the point of view of the audience of your updates. If the material you share is lively and exciting, your stakeholders will see the value in it and will look to each update with anticipation.

You don't necessarily need to *spin* things optimistically, but staying positive makes for enjoyable consumption of the content. Ultimately, you are aiming to build a *tribe*, to coin a phrase, and have them eagerly following your project.

Some techniques to consider for sharing project content include:

- project wikis (*covered previously in Chapter 16*);
- newsletters;
- blog posts;
- videos.

In addition to pushing out content in these forms, you can invite people to come along and observe project activities.

Invitations can be provided from time-to-time for stakeholders to sit-in and observe various project administration events such as sprint planning, prioritisation, demonstrations, etc.

This sort of transparent, behind-the-scenes approach really engages people.

Sprint Demonstrations

Sprint demonstrations, or demos, are one of the core activities in Scrum. They are held regularly at the end of each sprint, so the team can show the outputs they have completed to the project stakeholders.

A side-benefit of sprint demonstrations is they give the team regular mini-goals to aim for. Every sprint, the team knows that they will have an audience of people eager to see what they have completed. This generally provides a self-motivating incentive.

Team members have pride in their work and knowing that people will be paying attention drives them to deliver.

Here is a technique I have used to run sprint demonstrations over the past decade or so. First, create the framework for demos by way of a bulk set of one-off tasks completed early in the project:

- Decide who to invite.
- Pick dates.
- Send invites.
- Create a runsheet template.

Then, as the project progresses, complete the following steps for each sprint:

- Prepare the runsheet.
- Be ready.
- Run the demo.

In the following sub-sections, I provide detail about each of these steps. First, we start with establishing the sprint demo framework.

Decide Who to Invite

I usually limit sprint demos to a small number of key participants. The invite list I normally recommend includes members of the project board and the product owner.

The reason to restrict the group like this is simply to keep the session focused at a broad strategic level. This audience, the project board and the product owner, are generally viewing the project in terms of target outcomes, long term benefits and overall management.

A separate sprint output review session is held in addition to this demo so there is ample opportunity for other stakeholders to provide feedback on the finer details of the outputs. Keeping the sessions separate in this way tends to help the project board and product owner think in terms of the administration roles they have responsibility for.

Pick Dates

As soon as you know the schedule for sprints on the project, pick the dates for each demo.

My recommendation is that you set demos to be held in the afternoon on the last day for each sprint. This way, it becomes a known fixed deadline for each sprint and provides consistency to the project team's schedule.

Send Invites

Given the target audience for demos, i.e. the project board and product owner, it is often challenging to find common gaps in their diaries. After all, they are usually busy managers and executives.

To account for this, and to prevent situations where demo invitees can't make it due to a schedule clash, I like to send invites out well ahead of time.

If possible, once you know the project schedule and target dates, send invites out for all demos for the entire project. Yes, this means you could feasibly be sending invites for sessions several months away, possibly even over a year in the future. However, it is better to have the time locked away rather than dealing with availability conflicts and struggling to get people to turn up.

Create a Runsheet Template

Decide on a standard format for the sprint demos and then create a template the team can re-use for each session.

A general format I recommend is for individual team members to speak about and show the output(s) they have completed. To keep it simple I encourage those presenting to follow a rough script:

- Provide some background and context for the output being demonstrated, i.e. what user stories it came from and what target outcome(s) it is aiming to support.
- Show how the output works with a real example; i.e. actually demonstrate the output by using it.
- Invite questions to clarify anything.

At this point the framework for sprint demos is in place and the project is ready to move ahead. Once the project progresses, we see the following activities repeated for each sprint.

Prepare the Runsheet

Ensure the team gets together in their own time each sprint to prepare the runsheet for the demo. This involves writing down who is going to present what and the order of events.

Depending on how diligent the team is at self-organising this activity, you can leave it up to them to arrange. Or, you can support them with simple memory joggers such as recurring calendar reminders or a simple entry in the task management system for each sprint.

Do whatever works best, but just make sure it doesn't get overlooked. A good demo is one that is planned ahead of time and that has a bit of order to it.

Be Ready

There is nothing better than a demo where the session is set-up beforehand and everyone is ready to go on time. On the contrary, there is nothing worse than a sprint demo where the team isn't prepared

Having members of a project board and the product owner sit around while a bunch of unorganised people scramble to set up a half-baked demo is embarrassing and highly unprofessional.

Treat everyone involved with respect by getting ready early and ensuring the demo space is setup on time. If the demo includes remote attendees, this includes ensuring the communication technology used is configured and working properly.

Run the Demo

At this stage, the prior preparation will have served everyone well and the demo session itself should be quite straightforward. Team members show everyone the outputs they have completed in the sprint. Attendees are invited to ask questions and clarify anything they are not sure of.

The session concludes with a follow-on invitation for everyone involved to provide feedback using the project's preferred approach. More on this last step in the upcoming section later in this chapter.

Output Reviews

Output reviews are equivalent to "the one ring that brings them all and in the darkness binds them"[xvi]. In other words, output reviews are the centrepiece of Scrum projects.

Output reviews are where you get the user reference group and SMEs together to go through the outputs in detail. In these sessions, the prime goal is for people to provide feedback on how well they think the outputs meet their needs and fit their purpose.

The format and recommended steps for running output reviews is largely the same as for demonstrations. In other words, you still follow the same approach:

- Decide who to invite.
- Pick dates.
- Send invites.
- Create a runsheet template.

Then, as the project progresses, complete the following steps for each sprint:

- Prepare the runsheet.
- Be ready.
- Run the output review.

The two key differences of output reviews compared to demos are:

- The invited audience is much broader. Aim to get a good coverage of all the types of stakeholders to the project.
- The sessions are not so much demonstrations, but rather detailed hands on experiences. Instead of the project team showing their outputs, you want the attendees themselves to personally use the outputs.

Once again, preparation is key to a successful review session. Think of them as facilitated workshops or tutorial sessions. You basically want to be giving the reviewers guidance about what to review and what you would like them to pay attention to.

Focus on the reviewers thinking about the original requirements and needs that led to the user stories. Recall the steps-to-demonstrate and the acceptance criteria back in the original user stories associated with the outputs you are demonstrating. Ask the reviewers to work with your project's preferred feedback approach, which we cover in the next section.

Managing Feedback

Ahh, feedback ☺. If all goes to plan, you are going to get lots of it in your agile project. Lots and lots of it.

If you don't take a structured and systematic approach to managing feedback on your project, it is highly likely it will get out of hand fast. It is unwise to take an ad-hoc approach to feedback on Agile projects. Stakeholders are an intuitive lot and they will rapidly pick up on your chaotic vibes. Trust me, it's not a good look and could limit your project's success.

The main elements of managing feedback effectively are as follows:

- Accept feedback at any time.
- Encourage self-service.
- Use the task tracking system.
- Acknowledge and respond.
- Listen and take action.

Accept Feedback at Any Time

Ensure stakeholders have an open platform to say whatever they feel, whenever they like.

Don't limit your stakeholders to set times for providing feedback. Their thoughts and ideas regarding the project outputs can happen anytime.

Encourage Self-Service

Since you are likely to get large amounts of feedback, at random times, it is unsuitable for members of the project team to act as the conduit for feedback.

It makes more sense for stakeholders to write down their feedback and submit it to the team in their own time.

A side-benefit of this approach is that you avoid misinterpretation at the point of recording the feedback. That is, the person providing the feedback is the one that writes it down and therefore they get the opportunity to say exactly what they mean.

Use the Task Tracking System

Treat feedback as a first-class citizen in the project's task management system. In other words, find a way to enter items into the task management system and identify them as feedback.

By integrating like this, it is possible for feedback to get equal treatment in terms of prioritisation and focus. Some feedback may end up getting a high priority resulting in the project team undertaking work in response. Such work takes up capacity in a sprint and therefore it should be considered alongside user stories.

Following on from the previous two sections, the most ideal scenario is that you have a mechanism for self-provision of feedback that is integrated into the task management system. An example of this is some type of online feedback form that can be filled out and submitted by a stakeholder. When the form is submitted, the details are captured and entered directly into the task management system.

Acknowledge and Respond

For each item of feedback that comes in, thank the person for their submission. Let them know what will be done with it and when they are likely to get a response.

When further work is undertaken by the team on an item of feedback, ensure the submitter is updated on progress.

Doing so gives confidence to stakeholders that their feedback is valued and that it will be considered. Not everything will end up being actioned immediately and people are generally OK with this.

What is appreciated, however, is that people's thoughts and ideas are acknowledged.

Listen and Take Action

If you were to sum up the entire philosophy of agile project management in one sentence, it would be the following.

Listen to stakeholder feedback and adjust project plans to deliver outputs that take action on that feedback.

This is without a doubt one of the more important aspects of agile projects. For each piece of feedback that comes in, triage it and then decide how to respond to it.

Feedback is worthless if it isn't used. Listening to stakeholders is the first step. Taking action on their feedback is where the real value is.

Incorporating feedback into the thinking associated with backlog prioritisation will greatly improve the overall results of the project.

First, your stakeholders will see that you are listening to them and updating plans accordingly. This will give them confidence to continue engaging with you. Additionally, they will develop or strengthen their trust in you. They will gradually open up and relax further. Honest and candid opinions will follow.

Progress on the Conference Project

Grab that notebook of yours and get ready. It's time to put some more ideas into practice.

Who will be invited to the reference groups for the conference project? Are there past attendees that can be approached? What about local suppliers with significant experience producing and running conferences?

What mechanisms do you think would be worthwhile using for sharing project content? Would videos and regular blog posts help generate buzz about the conference? Would using a wiki help coordinate content for the people internal to the project?

How about the sprint demonstrations? Who should be invited? When and where will they occur? Since the conference is very much a form of *show*, would it be worth running some of them as *rehearsals* or *walkthroughs*.

In what way will you manage feedback for the conference project. Do you expect there to be much? If so, will you use some sort of system to help you deal with it?

It is often hard to see while it is happening, but I can guarantee you, based on several experiences, being open and inclusive in terms of progress greatly enhances the end results of the project in the long run.

A logical extension to the spirit of openness and inclusiveness is being transparent. In the next chapter, we consider status reporting for the project.

Chapter 21
Be Transparent

In this chapter, we cover status reporting on your project. It's time to bare it all.

We start with the basics of keeping people up-to-date with the relevant facts. We discuss some the reasons for reporting and what details to include. Additionally, we offer guidance on how often to report. I provide a template you can use to help produce your own status reports based on the ideas suggested here.

Following that, we consider the importance of remaining open, objective and honest on your projects and the benefits of doing so.

Status Reporting

Status reporting is the act of regularly communicating how the project is tracking against its plan. Remember, even though agile projects don't follow that old predictive-planning approach, there is still a plan in place.

Why Report?

For Awareness

Status reporting helps keep people informed about relevant facts, progress, risks, issues and any other important aspect of the project. The information provided in status reports helps everyone involved or interested in the project be aware of what is going on. It lets them know if there is any new information relevant to them.

To Alert Others

An agile project is a living, breathing beast. Most of the time, it will be growing organically toward its original goals, according to the plan.

Sometimes, things won't go to plan. There will be issues, that is simply a fact of life. Projects are unique, one-time activities often producing highly complex outputs. It is reasonable to expect issues from time to time.

When issues do occur, alerting others helps ensure everyone involved can respond accordingly. Some issues need others to help with treatment and getting the project back on track toward its plan.

To Celebrate

Status reporting isn't all about bad news. Good status reports are also a way to enjoy and acknowledge the achievements of everyone involved. Why not stay positive and inject a bit of life into work occasionally.

For Simplifying the Flow of Information

As the number of people on a project grows, the lines of communication between everyone increases exponentially. It rapidly becomes impossible for people to communicate effectively if they need to do it one-on-one.

Providing centralised status reports is the most efficient way of sharing information among everyone involved in a project. It becomes a consistent source of knowledge that can be accessed on-demand rather than having people go searching.

To Show Leadership

Status reporting is a great way to remind everyone there is a plan and that there is solid and confident leadership behind that plan.

In the absence of good status reporting, people involved in projects tend to wonder what is going on. This creates uncertainty since various narratives regarding progress begin to emerge.

Status reporting provides the source of truth that people can trust and rely upon.

To Help People Make Decisions

One of the more important reasons to produce regular status reports is to help others make informed decisions.

Projects don't exist in isolation. They are normally just one activity among a big set of others both internal and external to the organisation running the project.

People regularly need to make decisions and take action based on events that occur within a project. Regular status reports provide information that helps people to carry out their responsibilities.

What to Report?

Even though we are regularly hosting prioritisation sessions, sprint demos and output reviews in our Scrum project, there are many more facets that need to be monitored. Status reporting is a great way to keep stakeholders informed about all the other aspects beyond those specific to agile project management.

Some of the more relevant items to regularly report include:

- general summary;
- key aspects;
- progress toward budget, schedule, scope and outcomes;
- milestone tracking;
- people matters;
- current work in-progress;
- recently completed work;
- issues;
- risks;
- change requests;
- decisions;
- other matters.

How to Report

Consistency, reliability and ease-of-use are the keys to effective project status reporting.

- Report regularly.
- Choose suitable distribution channels.
- Use the same format and layout.

The goal is to make it as easy and reliable as possible for the people consuming the status reports.

Report Regularly

Pick a time interval that suits your project and the information needs of your stakeholders. It could be weekly, fortnightly, monthly. Once you have chosen a reporting timeframe, stick to it.

Choose Suitable Distribution Channels

Once again, depending on the preferences of the people involved in your project, pick a mechanism for sharing your status reports.

I tend to favour the combination of push notifications and self-serve consumption.

The way I do this is to send out an update to a list of recipients, usually via email. In the message, I provide a few key points and then let them know that the rest of the details are available from the project wiki. On the wiki, I provide the full status report that can be accessed any time.

Using this approach means that all status reports can be found together in the one place for the duration of the project. It also provides simple prompts for people to check-in on progress, without getting in their face.

Format and Layout

Work with your organisation and stakeholders to select a format and layout for your status reports that suits their needs. You could post directly to a project wiki.

Your organisation might alternatively have its own preferred reporting system that you enter information into.

On my projects, I tend to favour using a template reporting document. This way I can control the composition, design and layout of the report.

I find that being proactive in this way reduces the effort of communicating a project's status. I share relevant information according to my own preferred schedule, rather than being reactive to other people's random information needs.

Download a copy of the Project Status Report template at https://miller.productions/agile-toolbox/

For each report, I simply fill in this template based on the details of the project for that timeframe. Once it is ready, I publish a PDF version and send notifications out to people to let them know it is available.

The Importance of Being Earnest

I have a philosophy when it comes to project status reporting.

You get the support you deserve based on the transparency of your reporting.

Provide your project stakeholders with open, honest and factual reporting information and in return they will provide you with genuine support.

For anyone on the fence about how much or how little information to include in their status reports, consider these two views.

1. only sharing the good news
2. sharing everything, including the not-so-good news

If you only include good news in your status reports, people will get a false expectation about how the project is tracking.

There is only so long you can keep up a charade like this before you get found out.

Most likely, something disastrous will happen causing the whole façade to come tumbling down. Or, someone will eventually highlight the issue and people's confidence in you and the project team will be decimated.

Think about that scenario from another point of view. Consider the leader of an organisation with responsibility for a project. Whenever issues get escalated, their role is to assist with resolving those issues.

Treating an issue is a lot simpler the earlier they know about it. That way, the leader has plenty of time to think it through, consider all the options and then make the decision they think best.

If an issue is hidden and the leader finds out about it via other means, they will be disadvantaged. They will basically have to scramble to do something, fast. That is not the most enviable position to be in.

Think about how that will reflect on the people reporting the project status. Disappointing would be an understatement.

It is highly likely that a leader would not have much enthusiasm for people that have a habit of surprising them with unexpected issues.

Next, what about sharing the bad news? Could this be detrimental to the people associated with an issue?

For some people, being associated with a problem genuinely worries them. They think about how it will reflect on their professional reputation and how it could affect their careers. However, hiding from bad news doesn't help anybody in the long run.

Reporting Tips to Consider

- Be accurate with relevant and complete information.
- Stay objective and factual.
- Report the good and the bad.
- Consider solution options when reporting issues.
- Anticipate questions and be prepared with understanding of the information you report.
- Keep it relevant; only report information that others need.

- Show respect for individuals by not singling people out.
- Respect politics and hierarchies by checking with relevant people before publishing potentially sensitive information.

Status of the Conference Project

You know the drill by now. It's time to have a go at producing a status report for the conference project. Start by using the template provided in this chapter.

Once again, feel free to make this up as you go along. It doesn't really matter what you write here, just that you do have a go at it

Imagine we are about half-way through the conference project. The team has completed quite a few outputs by now and they are currently working on the latest set from the top of the project.

First, what is the general summary for the project. Are things going to plan? Or, is there currently a problem that people need to be aware of? Keep this section very short. One sentence only.

What are the highlights in the last reporting period? Has the team achieved any goals? Is there a new or novel idea? Did we overcome any challenges?

How about those constraints? Are we tracking over, under or on time? Same question for the budget. What about the scope? Are we still on track to match the general expectations?

For any milestones that we previously set for the project, what is the status against each of them? Are we on track to meet them? Are some at risk? Have we missed any?

Regarding the people on the project, is there anything we can share publicly?

For example, is anyone planning to go on a break/holiday in the near future? Do we have any new staff starting?

Next, we visit the work for the current reporting period. List out each of the items the team is currently working on.

Additionally, list the outputs they have recently completed. Are there any items that need highlighting? That is, are there any potential problems or things that need urgent attention?

Time to move on to the issues section. Are there any live issues that everyone needs to be aware of?

Only consider issues here that need attention from others. Don't include low-level trivial issues that the team can take care of without much hassle. Which issues are still open? Have we closed any recently?

They say that project management is basically the art of managing risks. Well, time to show off your artistic flair. In the risk section, list out the top risks for the project based on their rating. I usually include the top five risks in this section, give or take. Additionally, have any new risks arisen recently? Have we been able to close any risks completely?

Following the risks, we get down to the last three blocks.

What is the status of change requests on the project? Are there any open under consideration? Have any recently completed processing?

After that, summarise any decisions made by key stakeholders on the project. Don't be shy here. Shared information about decisions helps everyone stay abreast of ideas and directions.

Lastly, include anything else that you need to share about the project, that didn't fit into any of the other categories.

A Word About Size of Status Reports

Producing a status report like mine does take a bit of time, half an hour or so. You may be tempted to try and fit your status reports onto just one page. I would discourage this since you focus more on design and layout of the report rather than the details within.

Remember, status reporting is how you share the true state of the project. There is no benefit to cutting information out for the sake of fitting into a constrained document format. Focus on what people need to know, that is where the value in reporting lies.

Sharing positive results with the same level of openness as acknowledging issues is the sign of a true professional. Everyone loves good news, that is a no-brainer. However, reporting things that don't go so well is way better for projects in the long run.

Honestly, the more mistakes a project uncovers, the better off everyone will be in terms of outputs and skills. Issues are opportunities to learn, grow and improve over time.

If you take this sort of philosophy on-board, the project is more likely to reach its goals. Speaking of reaching goals, now is a good time to consider what happens when we reach the end of a project.

Chapter 22
Wrap it Up

Wow, can you believe it? The end is in sight! You have successfully planned and implemented your agile project. All those issues are behind you, the team is close to delivering the final set of outputs and you're on the home stretch. Well done.

Before we get ahead of ourselves, however, here is a sobering thought. We're nowhere near ready to down the tools, head for drinks and then go on holiday. There is still a bit of work to do.

In this chapter, we consider how to close an agile project. In addition to the usual suspects of handover and celebration, we look at stabilising outputs, concluding sprints, acknowledging completion with stakeholders and transitioning out of project mode into operations.

Include Closure in Your Plans

Closing an agile project takes time and effort. You can't expect everyone involved to keep producing and creating right up to the end of the project and then walk away.

Like any job worth doing, an agile job is worth doing well. This includes completing the tasks to finish up the work, check everything is in order, hand things over, archive materials and officially close. Once all this is finished, then everyone in the project can kick back and celebrate.

Whoa Back, Easy Does It

Stabilisation and Finishing Touches

Whenever you are continuously building a set of outputs from scratch, as is the case with agile projects, things are a bit unstable for a while. It takes time for all outputs in scope to be finalised to a point where they can all be considered *done.*

This has as much to do with the expectations of stakeholders as it does the skills of the team members responsible for output production.

Sometimes, the idea of done is conceptual and relates more to stakeholders' feelings about the functionality of an output. That is, accepting that what has been produced is all there is for now. In other words, coming to terms with the fact that the project doesn't have any further time or budget to keep going.

Other times, the definition of done is more tangible. For example, this can include times when project team members come back to complete everything in scope for the outputs they have been working on. There might be times where team members have left something for a while and then eventually come back to complete it.

You are going to need time at the end of an agile project for stabilisation and finishing touches of all the outputs in scope. A rule of thumb I follow is to allow at least one iteration or sprint for this.

Bug Fixing

Let's face it, nothing works perfectly the first time. When you produce anything from scratch, bugs are to be expected.

Nothing is more certain in life than death, taxes and bugs in new products.

Depending on how you treated bugs throughout your project, there may be a number leftover at the end. A simple approach in dealing with the leftover bugs is to divide them into at least two categories:

- bugs that must be fixed before the project is complete
- bugs that can be lived with for now and resolved at some later point in the future

For the bugs that must be fixed before the project is complete, now is the time to resolve them. I also include this activity into the final sprint on Scrum projects.

Stocktake

A common occurrence in agile projects is that the stakeholders usually want, or anticipate, more outputs than the team has capacity to produce. This is quite normal and to be expected.

Once the project is heading toward completion, it is time to undertake a final stocktake.

- List everything that was originally desired in the project. Sources of information for this include original project initiation and planning documents, notes from sprint planning and notes from backlog prioritisation sessions.

- List everything that has been completed by the project team. Sources of this information include sprint reports, project status reports, notes from sprint demonstrations, notes from output reviews and details contained in the project task tracking system.

- Cross-check the two lists against each other.

- Report the differences.

The most common difference will be items that were targeted but not completed. These shouldn't be a shock if you have been tracking the team's throughput rate at the end of each sprint.

Another difference you may see is additional items that were not expected initially. Hopefully, these can be explained by decisions made during prioritisation and output reviews.

If you find that items have been produced without a logical explanation, then this might be worth investigating further. Remember, the goal of agile is for the team to only produce items that the stakeholders have prioritised.

Handover

Once the work of creating, stabilising and fixing up the outputs of the project is complete, things are ready to be formally handed over.

This activity involves transitioning the outputs out of *project* mode and into *operations* mode.

It also involves transferring responsibility for the outputs from the project team, product owner and other key stakeholders of the project. It is usually the case that there will be a nominated individual or group of people that will be responsible for outputs post-project. Who this is depends on the context of the project, the organisation and the type of outputs produced.

You are aiming for conditions such as the following:

- The outputs are ready for use in their final production operational environment.
- There is sufficient capacity in terms of the people responsible for operating the outputs.
- The people responsible for the outputs have sufficient ability and understand how to use the outputs.
- The operations people are prepared to train themselves further, if necessary.
- There is sufficient documentation the people can refer to if necessary when maintaining the outputs.
- Arrangements are in place for post-project support of the outputs, if necessary.

Basically, make sure that conditions are optimal for the outputs to be utilised successfully by the end users.

Do whatever necessary here to help. Remember, successful projects deliver outcomes and benefits, not just outputs.

Sign-Off

This can be as formal or informal as you like. Again, it depends on the context in which your project is running. It also depends on how much internal vs external support there is on the project.

For purely internal projects, i.e. those where everyone involved is part of the same organisation, it is probably sensible for each party involved to simply acknowledge completion informally.

When there are organisations involved that are separate entities from each other, it is wise to formally conclude project activities with legally binding acknowledgements.

Either way, you are aiming for everyone to come to a shared conclusion:

- The project is now complete.
- The delivered outputs have been accepted.
- Responsibility has transitioned from the project to operations.
- Any issues and risks that remain open have been acknowledged.

Pack Up and Party

All good things must come to an end, sadly. As fun as agile projects are, there comes a time when the project is over. Kaput. Finito. At this stage, the work is truly done. It's time to pack up the tools, shut up the shop and hand any resources back that are no longer needed.

Following that, we can crank up the music, blow up the balloons and turn down the lights. Pick something that suits your organisation in terms of celebration and go for it! You've earned it! Project completion celebrations acknowledge the efforts of everyone involved. They are also a way for people who were working together to say farewell and get closure on that phase of their life.

While you're relaxing, there are a few last points I think you should consider…

PART IV
FINAL THOUGHTS

Chapter 23
It's Not All Sunshine and Rainbows

At this point, consider yourself enlightened about agile and ready to go forth into the world applying your new knowledge. Before you do, let's get something straight.

Agile is not *The Solution* to all your project management woes. Agile is not the magic project management pill that will work for every scenario.

It is a good idea, whenever learning a new technique, to also consider the risks so you don't end up hurting yourself or others around you. And believe me, running projects using agile certainly has its own fair share of risks.

Things can, and do, go wrong. I have seen and heard about several examples including the following:

- A company director that really wanted to try agile but wouldn't commit to providing funds for team members with sufficient skills and capabilities.
- Team members that thought they were being agile by working on whatever they personally felt would be best for the project rather than what stakeholders required.
- The organisation that was a tad confused, thinking they were running agile projects when instead they spent their time and effort on producing project management documentation and never really producing actual outputs.

- A project stakeholder group that were somewhat uninterested and didn't take the time to get involved in evaluating the project's outputs.
- The Chief that would panic each time the priorities of the project were reviewed and adjusted thinking it was scope creep or lack of good planning.
- The group that did almost everything right but tended to use agile as an excuse to not design, not document anything and not communicate their ideas back to their stakeholders. This group worked predominantly *by-the-seat-of-their pants*, i.e. figured things out as they went, which resulted in some outputs being wonderful in their own right but a really poor-fit to what was actually required.

In all these cases, and surely many others like them, the common factors are a lack of understanding on the part of the people running the project regarding:

- what agile is;
- what is required for agile to work well;
- what their individual roles are in the process.

Let's look at some of the more common issues that can arise in agile projects for even the most well-meaning teams with the best intentions.

Beware the Buzzword Name-Dropper

A colleague once told me about a situation they experienced that I'm sure we're all familiar with. A middle management team kept telling everyone their projects were agile when in fact things were being run just short of ad-hoc. Some of the managers just loved playing buzzword bingo and it seemed the term *agile* was the latest to catch their interest.

You'll occasionally encounter situations like this where a professional approach or concept is applied *in-name-only*. Pay close attention next time you come across someone mentioning they run agile projects. There could be three possibilities, maybe even more:

1. They are running their project using agile, possibly well.
2. They know they're not applying agile, but they want people to think they are keeping up with modern approaches.
3. They truly believe they are applying agile, but unfortunately are accidentally missing a few crucial aspects which results in issues detrimental to the project's success.

Our Own Flavour, Customised to Our Environment

While not exactly a bad thing, it is worth being wary of an organisation that says they run their own version of a professional approach. Usually this means a few dominant individuals have called the shots and the organisation has ended up with their own subjective opinion on running things. Agile is not immune to this.

Whether their own custom approach is successful or not depends on the individual(s) that came up with it in the first place. If they know what they are doing, you might be alright. If not, there could be much room for improvement.

Micro-Management

If you have experienced it, you know what I'm talking about. Nobody likes to be nit-picked. People certainly don't produce their best work if they are continuously being judged and corrected.

This really isn't unique to agile projects. However, due to the structured nature of agile there is a good chance that micro-management will show up. I have seen it occur both in managers and team members. It seems that whenever there are rules in a game or instructions for a way of doing business, someone always feels the need to tell everyone else what to do.

I won't go into a long-winded discussion about micro-management here. You could dedicate whole books about the topic. Instead, simply remember that one of the key principles of successful agile project management is self-organising teams. When you bring a team of motivated, experienced professionals together on a project they will give you their best when you believe in them. Show that you trust them to deliver and they will respect that trust.

If you are guilty of micro-management yourself, it is worth reflecting on what things trigger this pattern to come out in your day-to-day work. Being able to recognise the onset of the behaviour is a positive step toward controlling it rather than letting it take over you and your colleagues.

Laissez-Faire Leadership

At the opposite end of the scale from micro-management is disengaged leadership. This is when leaders of an agile project appear uninterested and don't really participate in the project at all.

In this scenario, leaders seem busy with several things. Unfortunately, your agile project is not one of their *things*.

They simply wave their hand and suggest that everyone else deal with that agile project.

The problem with this type of behaviour in leaders is that it permeates down throughout the rest of the organisation. Initially, you see executive managers not turning up to project activities. Then it begins to flow through to the stakeholders. Eventually, even the team starts losing motivation. Who would want to be involved in a project where nobody is interested?

To respond to this type of poor leadership, a pro-active approach is recommended. Spread enthusiasm for the project back up to the leaders. Let them know how valuable and important the project is. Celebrate any successes that the agile approach provides. Encourage active engagement wherever possible. It is challenging, but if you remain courteous, positive and professional at all times, eventually you might just find the leaders taking notice.

The Wrong Suit for the Occasion

Despite their best intentions, some organisations try to apply agile management to projects that are just not suited to it. Refer to Chapter 3 for a detailed look as to when this might occur.

Agile isn't something you can just have a go at. Care and deliberate evaluation is necessary to support decisions about whether to run projects using agile. In the wrong hands, agile is a tool that could do more damage than good.

Too Much Focus on the Tools and Techniques

Remember Chapter 4 and the overall theme of this book. People are the key to successful projects.

You occasionally see personality types that focus on the details of agile tools and techniques. This usually coincides with a disregard for the feelings of people, whether it is deliberate, accidental or oblivious.

Regardless, the impact is predictable. Agile projects where this occurs have very well defined and configured tools, techniques, processes and policies. However, they fail to deliver effectively because not enough time is spent listening to what people need.

Drifting Off-Topic

One notable aspect of agile projects is the number of meetings that are part of the process. This is particularly true of Scrum with its backlog prioritisation, sprint planning, demos, reviews, daily stand-ups and retrospectives.

Meetings are a major source of lost productivity in organisations, and a well-known cause of gripe for team members.

To achieve the benefits that are possible with agile processes, it is worth being focussed and efficient in meetings.

Nip those side conversations in the bud fast before everyone starts drifting off to discuss their plans for the weekend or their favourite holiday destinations. Protect everyone from trips down memory lane by reminding people to stay on topic.

Occasionally, there are valid conversations that come up during meetings. People begin discussing a topic and want to go further into the details.

When this happens, politely ask the participants to hold off the discussion until the end of a session. Once the main agenda items are complete, invite anyone that is interested or relevant to the emerging discussion to stick around and complete it.

The No Show

The rituals and meetings of agile are fundamental to the process. Particularly in Scrum, these formalities are deliberately designed to bring people together and encourage communication.

Every now and again, you get a situation where people outside the project team simply don't turn up to sessions. It is completely understandable: sometimes life gets busy.

However, the impact is quite high. Events are what makes agile projects work. When people go missing, the progress effectively slows down. There is a risk that things could even grind to a halt depending on the roles of absent people.

The two notable impacts of no shows are as follows:

- Team morale suffers since people begin to wonder if their effort is worth it if nobody can be bothered joining in.
- The connection between outputs and real feedback is severed and hence team members need to start making their own assumptions about what to build.

Making Assumptions

One of the main drivers of agile is producing outputs iteratively based on reviews and feedback of key stakeholders. Without this, you may as well just do-away with the formality of agile altogether.

Unfortunately, you occasionally see lapses in judgement or honest mistakes where team members forget to involve key stakeholders. Instead, team members default to the simplest approach of designing and producing solutions based on their own assumptions.

You can see why this would happen. A skilled professional can easily visualise a solution and rapidly get working on it. It takes much less effort to get on with building rather than having numerous conversations with key stakeholders about what they want.

The impact is that you are basically holding a lottery regarding target outcomes and benefits. Once the project is complete, the outputs might be a good fit. But then again, they might not. It depends how correct the assumptions are.

Over-Estimation

Estimation is hard. I can guarantee you that on every project, the team members will over-estimate their abilities at least once. That is, during the sprint planning process, the team will estimate that they can produce more than they end up completing.

It is human nature to under-estimate the complexity of producing outputs. "How hard could it be?" is the catch-phrase often heard or implied during estimation and planning sessions.

Pay attention to over-estimation on agile projects. It is a wicked problem with dangerous side-effects.

The more that a team over-estimates, the higher the expectation of stakeholders on the project. People begin to presume they will get more than what ends up being delivered. The flow on effects are:

- People are shocked later in the project when they find out they will get less than what they thought.
- Confidence in the ability of the team suffers and people start doubting them.
- Team morale drops since they never seem to achieve their goals and they regularly fail to deliver on commitments.
- Less important outputs get prioritised before others since stakeholders think they have a luxury of time and don't pay attention to items lower in the backlog.

Over-Engineering

Less common than over-estimation, but still an issue that comes up from time-to-time is over-engineering. This is where a project team produces more than what they need to. It could be outputs that are a higher quality than necessary to satisfy requirements. It could also be the production of outputs that provide little value.

One way to spot over-engineering is to look for outputs that are *behind-the-scenes*. That is, outputs that are not actively used by the customers. Such outputs are often described by team members as necessary or important. However, these opinions are in relation to the team members themselves and not the end users.

On agile projects, always remind yourself and the team that there is a fixed budget and schedule available. You can't produce everything you would like to. Each time the team works on one thing, it means other items from the backlog won't be worked on.

Anytime you suspect over-engineering, consider if there are any other outputs with a higher value that could be worked on instead.

Half-Baked, Not Done

In the rush to get through as many outputs as possible in a sprint, teams occasionally don't *fully complete* items they work on.

Statements are made along the lines of "I'll come back to this later". The truth is, people never really have the time to come back and finish those things they leave behind. Something more important or interesting always comes up.

The impact is that you end up with a project littered with half-baked, not done outputs. Then, at the end of the project, people must scramble to tidy things up and get everything ready for final delivery.

It is better for the team to be honest with their abilities and estimates by completing each output they work on to a satisfactory level. Sure, this usually takes more time and effort, but the end results are far superior. A project that delivers a smaller number of fully complete outputs is better than one with a larger number of incomplete items.

There Are More

The issues included in this chapter are just some of the main ones that come up often and have the highest impact. There are others out there, but rather than dwelling on them I instead offer the following concluding advice. Agile is like any tool. In the right hands it works very well, but a fool with a tool is still a fool.

Think carefully about how *you* apply agile on *your* project.

With that final sentence in mind, I bid you adieu. Now get out there and have a bit of fun with your agile projects. They'll run smoother now that you know how to keep the herd comfortable.

Acknowledgements

1. Associate Professor Ofer Zwikael & John Smyrk

https://researchers.anu.edu.au/researchers/zwikael-oy

http://www.smscience.com/

Some of the material in Chapter 4 as well as the references to target outcomes and benefits is attributable back to the original work of Associate Professor Ofer Zwikael and John Smyrk in their book *Project Management for the Creation of Organisational Value*[xvii]. I encourage you to purchase a copy as a companion to this one.

With several decades of combined research, teaching and application between them in the field of project management, Ofer and John provide an excellent model for the roles associated with projects. In addition, they are known leaders in the areas of outcomes and benefits, the true measures of project success.

In the early years of my career in project management I was consuming as much documentation, textbook material and general reading as I could to build up my knowledge. As you could guess, I encountered many definitions for the roles associated with projects. My conclusion is that the work of Ofer and John makes the most logical sense. Their material is pragmatic and easily applied to real scenarios that you see in practice.

Gentlemen, thank you. You both inspired me to advance my project management career and have provided me with an enduring respect for the profession.

2. Professor Walter Fernandez

https://www.business.unsw.edu.au/our-people/walter-fernandez

Walter was one of the first people to teach me about project management during my university days. I enrolled in Walter's class while studying for my MBA at the Australian National University.

Through Walter, I learned the importance of socio-cultural factors when it comes to managing projects. He also gave me my first big break at tutoring post-grad students. This eventually led to me advancing to become a lecturer and from that point forward my drive and enthusiasm for project management was set in motion.

Thank you, Walter, you are true leader.

3. Staff at the Commonwealth Science & Industrial Research Organisation (CSIRO)

http://csiro.au/

Much of what I have learned and applied regarding agile project management has been made possible by the support of numerous staff at Australia's CSIRO.

Some of the most enjoyable projects I have managed in my career to date have been with the CSIRO.

Everyone I worked with in the organisation, and this goes without exception, enabled success for the projects I was responsible for. Their unwavering dedication, professionalism and all-round positive approach to work is truly inspiring. It is no wonder the folks at CSIRO have such a positive impact on the world.

To each and every person I have ever had the privilege of working with at CSIRO, thank you.

4. Stephen Mellor

http://stephenmellor.com/

Stephen is one of the signatories to the Agile Manifesto.

I had the pleasure of being taught by Stephen while undertaking post-graduate study at the Australian National University. Stephen was a guest lecturer in the Master of Software Engineering program.

Stephen was one of the first people that introduced me to the concepts of agile project management and the Agile Manifesto. My experience with Stephen was that he was a wonderful teacher and he really opened my eyes up to new possibilities in running projects.

5. Image Credits

Audience, p137, by Free-Photos at Pixabay
https://pixabay.com/en/audience-crowd-people-persons-828584/

Deer image, front cover, by Laura College on Unsplash
https://unsplash.com/photos/K_Na5gCmh38

Deer icon, spine and inside cover, by Tae S Yang from Flaticon
https://www.flaticon.com/free-icon/deer_130647

Endnotes

[i] Oxford Dictionary, definition of Agile in English, accessed 17/08/2017, https://en.oxforddictionaries.com/definition/agile
[ii] *Manifesto for Agile Software Development* (2001), accessed 17/08/2017, http://agilemanifesto.org/
[iii] Bernie Thompson (2008). *Scrum-ban*. Lean Software Engineering, Essays on the Continuous Delivery of High Quality Information Systems. http://leansoftwareengineering.com/ksse/scrum-ban/, accessed 17/08/2017
[iv] Alistair Cockburn (2004). *Crystal Clear: A Human-Powered Methodology for Small Teams: A Human-Powered Methodology for Small Teams*. Addison-Wesley Professional, Pearson Education, NJ, USA.
[v] Ambler, Scott. *The Agile Unified Process (AUP)*. Ambysoft. http://www.ambysoft.com/unifiedprocess/agileUP.html, accessed 16/08/2017.
[vi] Scott Ambler and Mark Lines (2012). *Disciplined Agile Delivery: A Practitioner's Guide to Agile Software Delivery in the Enterprise*. IBM Press.
[vii] Mary Poppendieck; Tom Poppendieck (2003). *Lean Software Development: An Agile Toolkit*. Addison-Wesley Professional.
[viii] Jim Highsmith (2002). *Agile Software Development Ecosystems*. Addison-Wesley Longman Publishing Co., Inc., Boston, MA, USA.
[ix] Agile Business Consortium (2014). *The DSDM Agile Project Framework (2014 Onwards) Handbook*. Dynamic Systems Development Method Limited, Kent, UK.
[x] VersionOne.*State of Agile Report*. http://stateofagile.versionone.com/, accessed 15/08/2017.

[xi] Scrum Alliance. *State of Scrum Report.* https://www.scrumalliance.org/why-scrum/state-of-scrum-report, accessed 15/08/2017.
[xii] Hobbs, B. & Petit, Y. (2017). *Agile Methods on Large Projects in Large Organizations.* Project Management Journal, 48(3), 3–19.
[xiii] Project Management Institute (PMI), *What is Project Management?* https://www.pmi.org/about/learn-about-pmi/what-is-project-management, accessed 28/07/2017.
[xiv] Doran, George T. *There's a S.M.A.R.T. way to write management's goals and objectives.* Management Review 70.11 (Nov. 1981): 35. Business Source Corporate.EBSCO. 15 Oct. 2008.
[xv] Wikipedia, *Comparison of project management software,* https://en.wikipedia.org/wiki/Comparison_of_project_management_software, accessed 28/07/2017.
[xvi] Excerpt from the Ring Verse. Tolkien, J.R.R. (1954). Lord of the Rings. 2nd Ed. England: Houghton Mifflin.
[xvii] Zwikael, O. & Smyrk, J. (2011). *Project Management for the Creation of Organisational Value.* 1st ed. London: Springer-Verlag.

Other work by Dan Miller

...visit https://miller.productions/

CPSIA information can be obtained
at www.ICGtesting.com
Printed in the USA
BVHW031753260120
570544BV00001B/14